Problems Book for Probabilistic Methods for the Theory of Structures with Complete Worked Through Solutions

Problems Book for Probabilistic Methods for the Theory of Structures with Complete Worked Through Solutions

Isaac Elishakoff
Distinguished Research Professor
Florida Atlantic University, USA
&
Visiting Distinguished Professor
Technion–Israel Institute of Technology, Israel

World Scientific

NEW JERSEY • LONDON • SINGAPORE • BEIJING • SHANGHAI • HONG KONG • TAIPEI • CHENNAI • TOKYO

Published by

World Scientific Publishing Co. Pte. Ltd.
5 Toh Tuck Link, Singapore 596224
USA office: 27 Warren Street, Suite 401-402, Hackensack, NJ 07601
UK office: 57 Shelton Street, Covent Garden, London WC2H 9HE

British Library Cataloguing-in-Publication Data
A catalogue record for this book is available from the British Library.

PROBLEMS BOOK FOR PROBABILISTIC METHODS FOR THE THEORY OF STRUCTURES WITH COMPLETE WORKED THROUGH SOLUTIONS

ISBN 978-981-3201-10-1
ISBN 978-981-3201-11-8 (pbk)

For any available supplementary material, please visit
http://www.worldscientific.com/worldscibooks/10.1142/10311#t=suppl

Typeset by Stallion Press
Email: enquiries@stallionpress.com

Printed in Singapore

Contents

Preface

The first edition of the combined monograph and textbook *Probabilistic Methods in the Theory of Structures* was published by Wiley-Interscience in 1983. 1n 1999, Dover Publications, Inc. published its second edition under shorter title *Probabilistic Theory of Structures.* Now, World Scientific kindly publishes, simultaneously, the third, corrected and expanded edition, and its companion volume containing *Problems with Complete Worked-Through Solutions.* This compendium of solutions was written in response to requests by numerous university educators around the world, since it has been adopted as a textbook or an additional reading both undergraduate and graduate courses.

I hope that the availability of such solutions manual will further help to establish the courses dealing with probabilistic strength of materials, design, random buckling, and random vibration. The material itself was developed and used by me during teaching various undergraduate and graduate courses at the Technion, during years 1972-1989, at the Delft University of Technology during the academic year 1979/80, and at the University of Notre Dame during the academic year 1985/86, and at the Florida Atlantic University since 1989.

Already since mid-eighties I was informed that the book was adopted in numerous universities worldwide. Besides complete solutions to more than one hundred problems, additional material and remarks are included as Chapter 12, bringing some ideas down to the "number" level. Errors and misprints (as a random event) are obviously inevitable. I hope that they are few and that readers will be kind enough to call my attention to them so that they can be corrected.

It is a pleasure to acknowledge the help I received during the preparation of the manual. Amongst my colleagues at the Aeronautical Department at the Technion-I.T.T., I am indebted to Dr. Yehuda Stavsky, Gerard Swope Professor in Mechanics, and Professor Menahem Baruch, then Dean of the Department, for discussions on number of probabilistic topics. The manual was received much boost during my stay at the Department of Aerospace and Mechanical Engineering of the University of Notre Dame, as a Visiting Frank M. Freimann Chair Professor, during the academic year 1985/1986. My sincere thanks are due to Professor Albin A. Szewzyk, Chairman of the Department, as well as the staff, for providing an ever pleasant and encouraging atmosphere.

Among my numerous students, my sincere thanks are due to Dr. Gabriel Cederbaum, currently Associate Professor at the University of Ben Gurion, Beer-Sheva, Israel, providing a number of solutions and painstakingly going through most of them. Credit for editing the text goes to my friend of long standing Mr. Eliezer Goldberg, Eng. Last, but not least, I wish to thank Mr. Brandon Naar of the Florida Atlantic University, for typing the entire manuscript in Fall 2015 and Spring 2016 semsters. I am very appreciative to Ing. Damien Delbecq of IFMA- French Institute of Advanced Mechanics for creating the figures, during his stay at the Florida Atlantic University during academic year 2015/6. Special thanks are due to Ms. Rochelle Kronzek, Executive Editor of the World Scientific Publishing Company for her indefatigable insistence this solutions manual to be submitted for publication so as to help the lecturer and student alike in understanding the nitty-gritty of probabilistic methods in structural applications.

It is strongly hoped that this manual will promote much wider dissemination of probabilistic methods' courses at universities, and ultimately, in engineering practice worldwide.

Isaac Elishakoff
Boca Raton, Florida, May, 2016

chapter 2

Probability Axioms

PROBLEM 2.1

Present a Venn diagram for $\bar{C}$, where $C = B \backslash A$.

SOLUTION 2.1

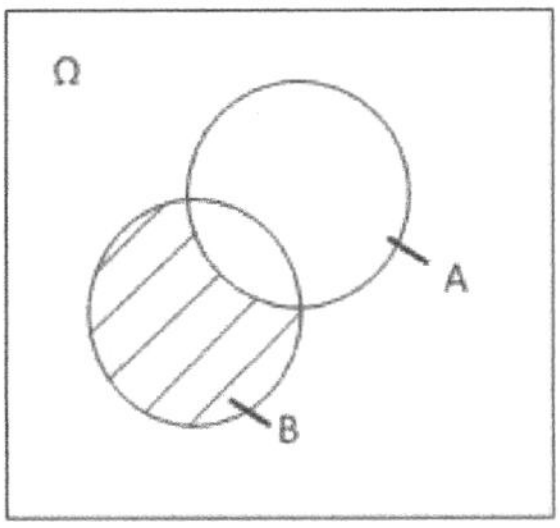

$C = B/A$ Hatched

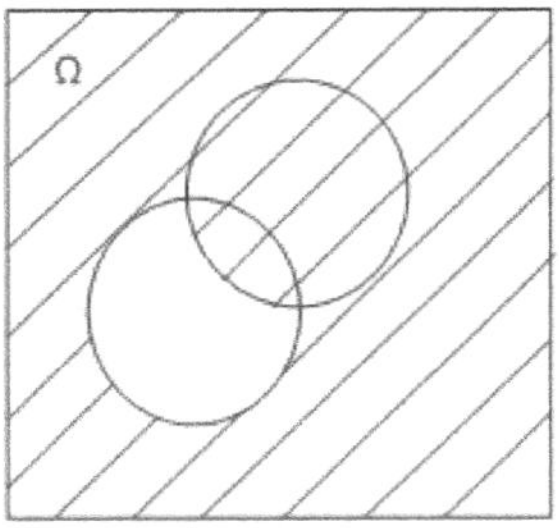

$\bar{C}$ hatched

A and B have common points

If A and B have no joint points

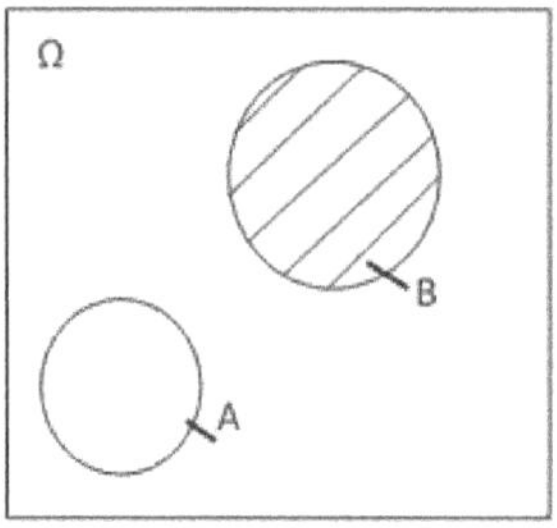

$C = B/A$ hatched

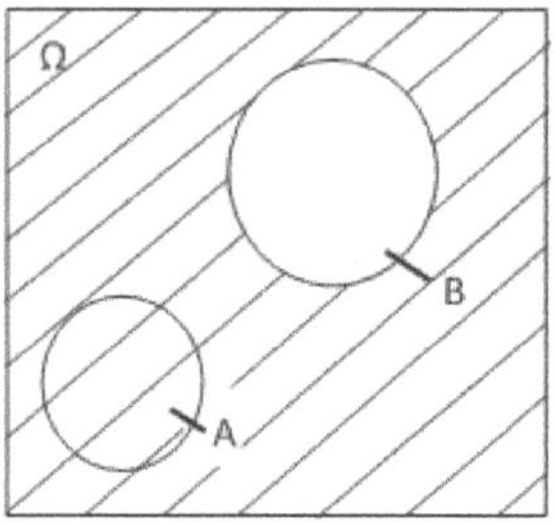

$\bar{C}$ hatched

$A \subset B$

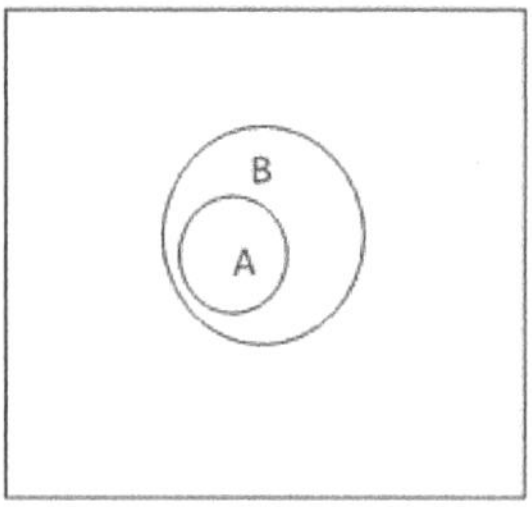

$A \subset B$

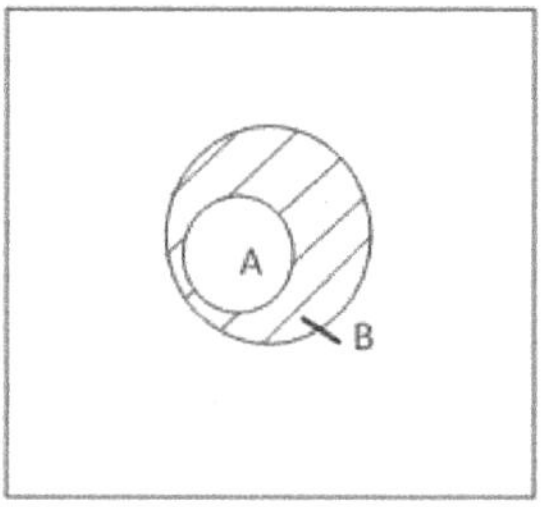

$C = B/A$ hatched

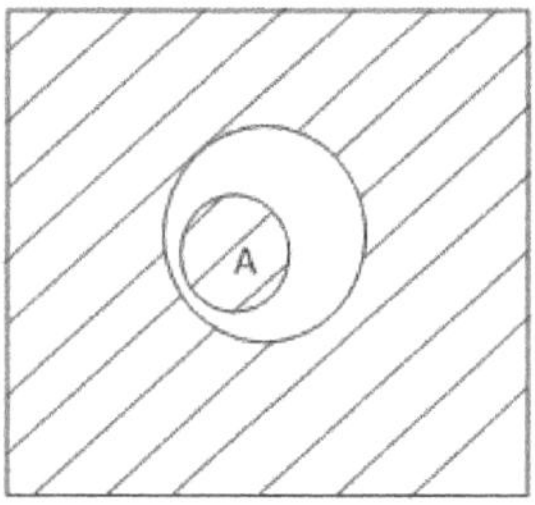

$\bar{C}$ hatched

PROBLEM 2.2

Verify by means of a Venn diagram that a union and an intersection of random events are distributive, that is,

$$(A \cup B) \cap C = (A \cap C) \cup (B \cap C)$$

$$(A \cap B) \cup C = (A \cup C) \cap (B \cup C)$$

SOLUTION 2.2

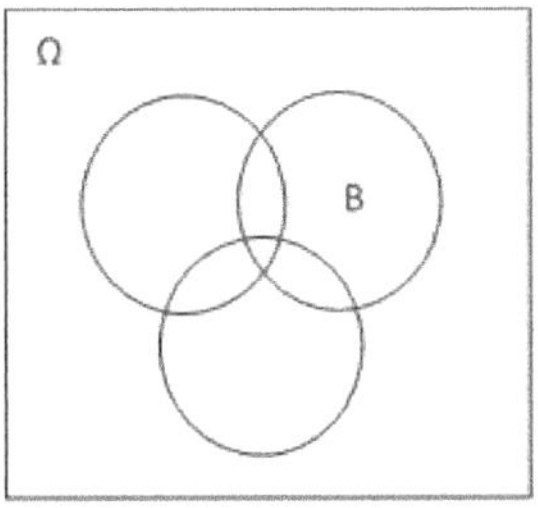

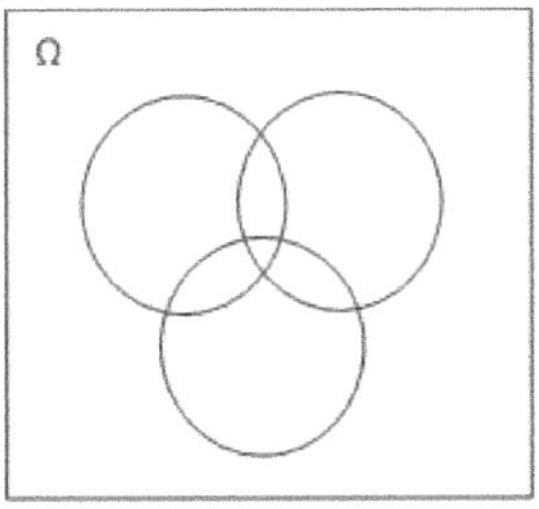

$A \cup B$

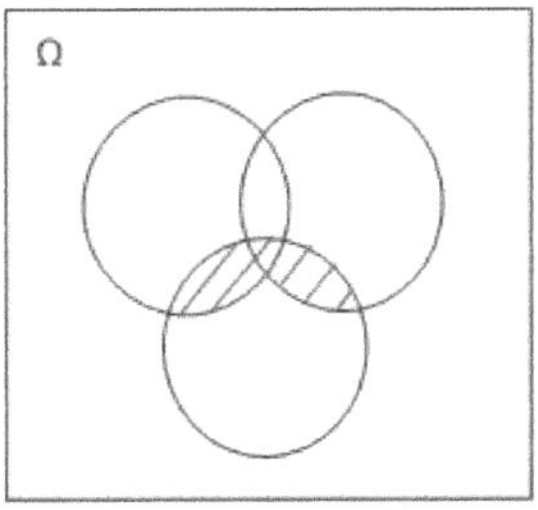

$(A \cup B) \cap C$

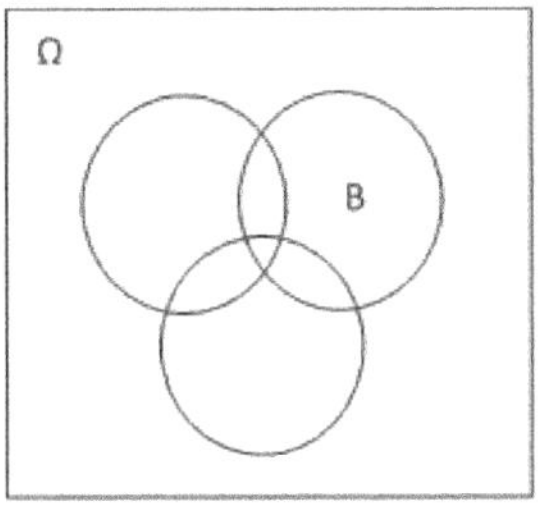

$A \cap C$

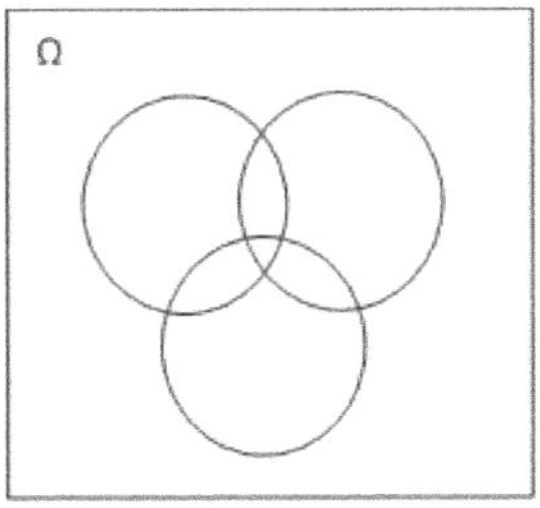

$B \cap C$

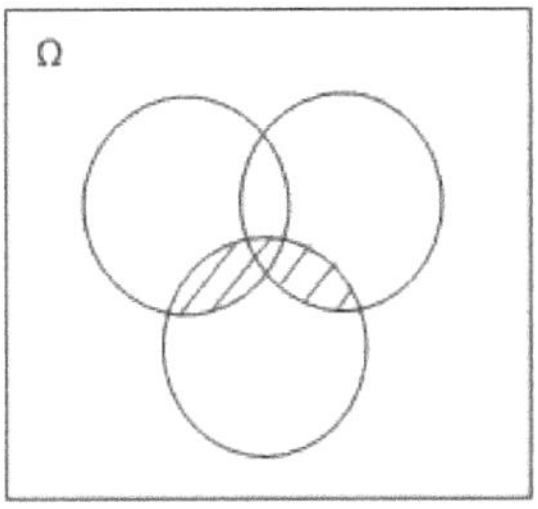

$(A \cap C) \cup (B \cap C)$

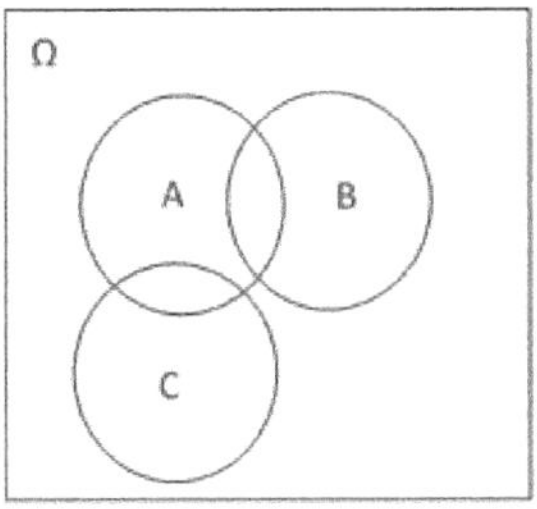

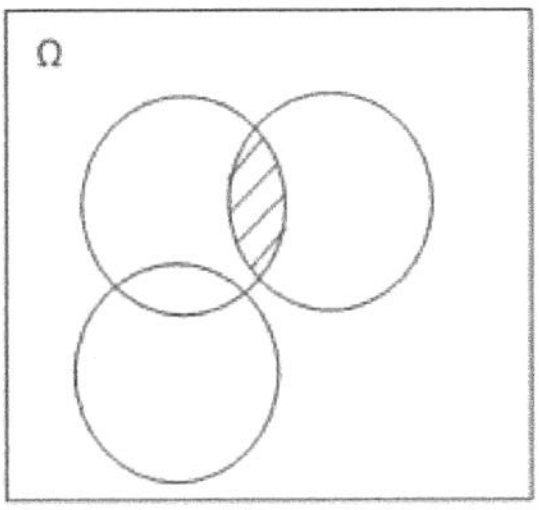

$A \cap B$

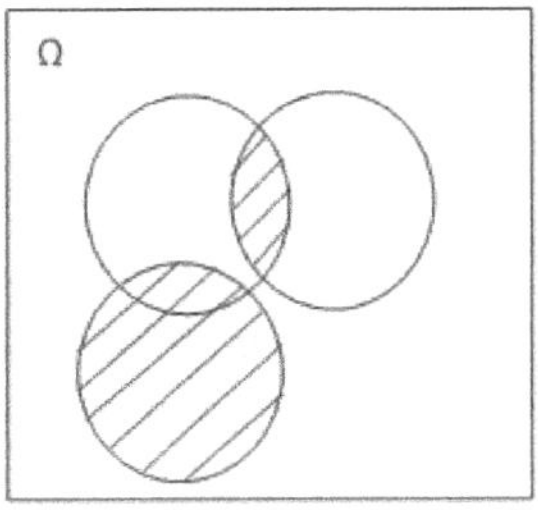

$(A \cap B) \cup C$

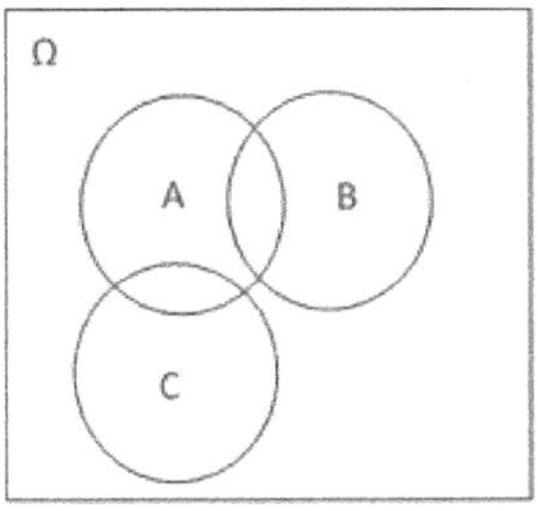

$A \cup C$

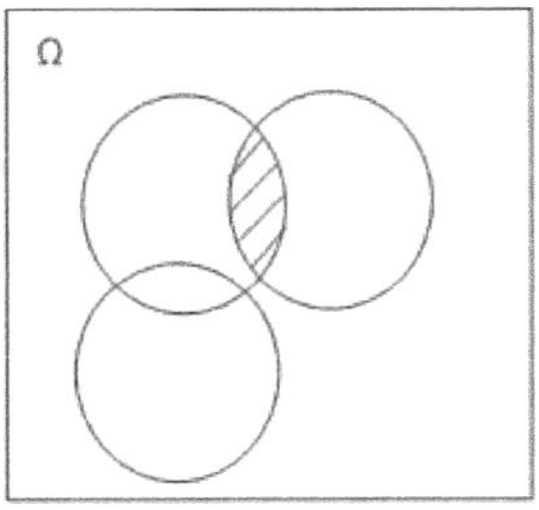

$B \cup C$

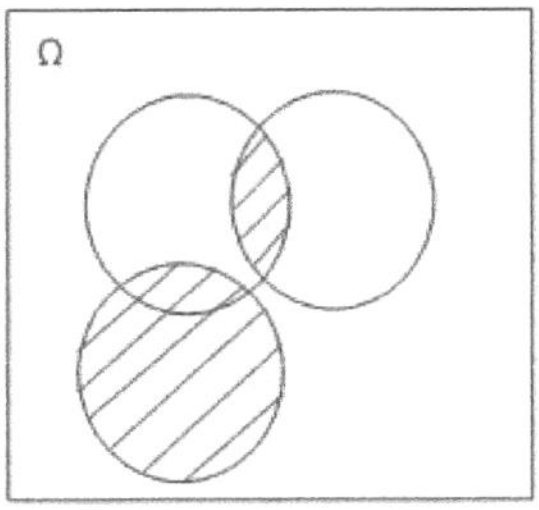

$(A \cup C) \cap (B \cup C)$

PROBLEM 2.3

A telephone relay satellite is known to have five malfunctioning channels out of 500 available. If a customer gets one of the malfunctioning channels on first dialing, what is the probability of his hitting on another malfunctioning channel on dialing again?

Note to Lecturer: Better to change "Another Malfunction Channel" to "Again Hitting on a Malfunctioning Channel."

SOLUTION 2.3

The random event (A_2) of hitting on a malfunctioning channel with second dialing is independent of what happened during previous dialings (A_1), therefore the probability of failure at the second dialing equals that at the first:

$$P(A_2|A_1) = P(A_2) = 0.01$$

The problem is equivalent to the following:

An opaque urn contains "a" white balls are "b" black balls. What is the probability of drawing a white ball in an honestly mixed lot in the course of Successive trials?

(1) The probability of drawing a white ball at the first trial is

$$P(A) = \frac{a}{a+b}$$

(2) The probability of drawing a white ball at the second trial depends on additional information: If the first ball is returned to the urn, then $P(B) = P(A) = \frac{a}{a+b}$. Such an experiment is designated "drawing" with replacement." In the original problem 2.3 the urn is analogous to the 500 channels; drawing a white ball is analogous to a hitting malfunctioning channel drawing a black ball into hitting a sound channel.

The channel hit on at the first dialing can be hit on again at the second. This is analogous to the possibility of drawing the same ball again in the urn experiment with replacement.

PROBLEM 2.4

Let m items be chosen at random from a lot containing $n > m$ items of which $p(m < p < n)$ are defective. Find the probability of all m items being nondefective. Consider also the particular case $m = 3$, $p = 4$, $n = 8$.

FIRST SOLUTION 2.4

The probability of choosing a nondefective item without returning (*WR*) is

$$P(A_1) = \frac{n - P}{n}$$

The probability of choosing another nondefective item, *WR* is:

$$P(A_2|A_1) = \frac{n - p - 1}{n - 1}$$

Therefore the probability of choosing two nondefective item, *WR* is:

$$P(A_1 A_2) = P(A_1)P(A_2)(A_1) = \frac{(n - p)(n - p - A)}{n(n - 1)}$$

The conditional probability of choosing the m-th nondefective item

$$P(A_m|A_1 A_2 \cdots A_{m-1}) = \frac{n - p - m + 1}{n - m + 1}$$

consequently, the probability of all items being nondefective equals

$$P(A_1 A_2 \cdots A_{m-1} A_m) = P(A_m|A_1 A_2 \cdots A_{m-1}) P(A_1 A_2 \cdots A_{m-1})$$
$$= \frac{(n - p)(n - p - 1) \cdots (n - p - m + 1)}{n(n - 1) \cdots (n - m + 1)}$$

For $m = 3$, $p = 4$, $n = 8$

$$P = \frac{4 \times 3 \times 2}{8 \times 7 \times 6} = \frac{1}{14}$$

SECOND SOLUTION 2.4

The number of combinations of m items that can be chosen from a lot containing n items is

$$\binom{n}{m} = \frac{n!}{(n - m)!m!}$$

Our interest is confined to nondefective items, i.e. m items are to be from a lot of $(n - p)$ nondefective items. The number of such combinations is

$$\binom{n - p}{m} = \frac{(n - p)!}{(n - p - m)!m!}$$

The sought probability is therefore

$$P(A) = \frac{\binom{n-p}{m}}{\binom{n}{m}} = \frac{(n-p)!}{(n-p-m)!m!}\frac{(n-m)!m!}{n!} = \frac{(n-m)!(n-p)!}{n!(n-p-m)!}$$

$$= \frac{(n-m)!}{n!}\frac{(n-p)!}{(n-p-m)!} = \frac{1}{(n-m+1)(n-m+2)\cdots n}$$

$$\times \frac{(n-p-m+1)(n-p-m+2)\cdots(n-p)}{1}$$

which is identical in the result by the first method of solution.

SIMPLER SOLUTION FOR A PARTICULAR CASE 2.4

The probability of choosing the first nondefective item is

$$P(A_1) = \frac{8-4}{8} = \frac{4}{8} = \frac{1}{2}$$

The probability of choosing the second nondefective item, provided the first was nondefective as well

$$P(A_2|A_1) = \frac{4-1}{8-1} = \frac{3}{7}$$

The probability of choosing the third nondefective item provided the first two are a nondefective

$$P(A_3|A_1A_2) = \frac{4-2}{8-2} = \frac{2}{6} = \frac{1}{3}$$

Finally, the sought probability of all three being nondefective is

$$P(A_1A_2A_3) = \frac{1}{2} \times \frac{3}{7} \times \frac{1}{3} = \frac{1}{14}$$

PROBLEM 2.5

A single playing card is picked at random from a well-shuffled ordinary deck of 52. Consider the events A, king picked; B, ace picked; C, heart picked. Check whether (1) A and B, (2) A and C, (3) B and C, are dependent or independent.

SOLUTION 2.5

$$P(A) = \frac{4}{52} = \frac{1}{13}$$

$$P(B) = \frac{4}{52} = \frac{1}{13}$$

$$P(C) = \frac{13}{52} = \frac{1}{4}$$

(1) Events A and B are dependent, being mutually exclusive. On either of them occurring, the probability of the other vanishes $P(A|B) = P(B|A) = 0$ or formulated deferently: $P(AB) = 0 \neq P(A|B)P(B)$

(2) $P(A|C) = \frac{1}{13} = P(A)$ and, moreover

$$P(C|A) = \frac{1}{4} = P(C)$$

Events A and C are independent.

Alternatively:

$$P(AC) = \frac{1}{52} = P(A)P(C) = \frac{1}{13} \cdot \frac{1}{4}$$

(3) As in (2):

$$P(B|C) = 13^{-4} = P(B)$$

$$P(C|B) = 4^{-4} = P(C)$$

B and C are independent, or alternatively:

$$P(BC) = \frac{1}{52} = P(B)P(C) = \frac{1}{13} \cdot \frac{1}{4}$$

PROBLEM 2.6

In a family of two children the older child is a boy. What is the probability of both of them being boys?

SOLUTION 2.6

We assume that a successfully delivered child is equally likely to be a boy or a girl, the attendant probability being one-half in either case. The sample space thus contains the following four elementary events with equal likelihood:

ELEM. EVENTS / CHILDREN	1	2	3	4
OLDER CHILD	Boy	Boy	Girl	Girl
YOUNGER CHILD	Boy	Girl	Boy	Girl

Out of these 4 events, two are favorable to be situation in question:

$$P(A) = \frac{2}{4} = \frac{1}{2}$$

ANOTHER SOLUTION 2.6

Formula (2.11) $P(A|B) = \frac{P(AB)}{P(B)}$ can be interpreted as follows: If we discard all events of the space in which event B did not occur, and retain the subspace (or a "new" sample space) in which it occurred, then $P(A|B)$ equals the relative frequency of occurrence of event A in that subspace. In our case we denote:

B– older child is a boy
A– both children are boys.

The sub space (or new sample spaces) contains two events

ELEM. EVENTS / CHILDREN	1	2
OLDER CHILD	Boy	Boy
YOUNGER CHILD	Boy	Girl

Here only the first is favorable, therefore:

$$P(A|B) = 1/2$$

or, directly as Eq. (2.11)

$$P(AB) = \frac{1}{4}, \; P(B) = \frac{2}{4}, \; P(A|B) = \frac{1/4}{2/4} = \frac{1}{2}$$

PROBLEM 2.7

In a family of two children, at least one of them is a boy. What is the probability of both of them being boys?

SOLUTION 2.7

The sample space is as in the preceding problem.

Denote:

A – one of the children is a boy
B – both children are boys

$$P(B|A) = \frac{P(AB)}{P(A)}$$

Now:

$$P(AB) = P(B) = \frac{1}{4}$$

$$P(A) = \frac{3}{4}$$

Therefore: $P(B|A) = \frac{1/4}{3/4} = \frac{1}{3}$

ANOTHER SOLUTION 2.7

Compare with the "other solution" of Prob. 2.6. The "new" sample space is

ELEM. EVENTS / CHILDREN	1	2	3
OLDER CHILD	Boy	Boy	Girl
YOUNGER CHILD	Boy	Girl	Boy

Only one of these equally likely events is favorable to the situation, therefore, the sought probability is $1/3$.

PROBLEM 2.8

A statically determinate truss consists of n bars, whose failures represent independent random variables with identical probabilities p. Find the permissible value of p such that the probability of failure Q of the entire truss does not exceed some prescribed value q.

SOLUTION 2.8

Denote A_i — failure of i-th bar. The reliability is then

$$R = P(\bar{A}_1 \bar{A}_2 \cdots \bar{A}_n)$$
$$= P(\bar{A}_1)\, P(\bar{A}_2) \cdots P(\bar{A}_n) = (1-p)^n$$

The probability of failure of the entire truss becomes

$$Q = 1 - (1-p)^n$$

The requirement that the probability of failure does not exceed the prescribed value γ leads to

$$1 - (1-p)^n \le q$$

complying

$$p \le 1 - \sqrt[n]{1-q}$$

For example, if $n = 7$ (like in Fig. 2.9a) and $q = 0.01$, $p \le 0.0014347$.

PROBLEM 2.9

A space shuttle is assigned to visit a TV satellite in a GSO (geosynchronous orbit) with radius 36,000 km (event A_1), and carry out maintenance (event B_1). Alternatively, it may, due to failure of one of one of the engines, go into an LEO (low Earth orbit) with radius 200 km (event A_2) and carry out a crop survey (event B_2). B_1 and B_2 are regarded as successful performance. The third possibility is failure to take off (event A_3). What is the reliability (probability of successful performance) of the space shuttle if $P(B_1 \mid A_1) = 0.75$, $P(B_2 \mid A_2) = 0.85$, $P(A_1) = 0.80$, and $P(A_2) = 0.15$?

SOLUTION 2.9

Successful performance of the space shuttle is defined as occurrence of either event B_1 or B_2. Reliability is therefore

$$R = P(B_1 \cup B_2)$$

But B_1 and B_2 are mutually exclusive, therefore

$$R = P(B_1) + P(B_2)$$

Now

$$P(B_1) = P(B_1A_1) = P(B_1 \mid A_1)\ P(A_1) = 0.75 \times 0.80 = 0.60$$
$$P(B_2) = P(B_2A_2) = P(B_2 \mid A_2)\ P(A_2) = 0.85 \times 0.15 = 0.1275$$

and

$$R = 0.6 + 0.1275 = 0.7275$$

PROBLEM 2.10

(Birger) During inspection of a gas turbine, two symptoms are checked: increase of the engine gas temperature at the turbine outlet by more than $50°C$ (symptom k_1) and increase of the acceleration time of the rpm from minimum to maximum, by more than $5s$ (symptom k_2). Assume that for the given type of engine these symptoms are associated either with failure of the fuel flow regulator (state D_1) or with reduction of the radial clearance of the turbine (state D_2). The normal state is denoted by D_3. The following probabilities are given:

$$P(D_1) = 0.05,\ P(k_1|D_1) = 0.2,\ P(k_2|D_1) = 0.3$$
$$P(D_2) = 0.15,\ P(k_1|D_2) = 0.4,\ P(k_2|D_2) = 0.5$$
$$P(D_3) = 0.80,\ P(k_1|D_3) = 0.0,\ P(k_2|D_3) = 0.05$$

Show that

$$P(D_1|k_1k_2) = 0.09,\ P(D_2|k_1k_2) = 0.91,\ P(D_3|k_1k_2) = 0.0$$
$$P(D_1|\bar{k}_1k_2) = 0.12,\ P(D_2|\bar{k}_1k_2) = 0.46,\ P(D_3|\bar{k}_1k_2) = 0.41$$
$$P(D_1|\bar{k}_1\bar{k}_2) = 0.03,\ P(D_2|\bar{k}_1\bar{k}_2) = 0.05,\ P(D_3|\bar{k}_1\bar{k}_2) = 0.92$$

indicating, for example, that in the presence of symptoms k_1 and k_2 the probability of reduction of the radial clearance is about ten times that of failure of the fuel flow regulator.

SOLUTION 2.10

To solve Birger's problem, we report to Bayes' formula (2.26) which reads:

$$P(A_i|B) = \frac{P(B|A_i)P(A_i)}{\sum_{i=1}^{3} P(B|A_i)P(A_i)} \tag{2.26}$$

Let us first calculate $P(D_i|k_1k_2)$

Formal substitution of $A_i \to D_i$, $B \to k_i \cap k_2 = k_1k_2$ into Eq. (2.26) yields:

$$P(D_i|k_1k_2) = \frac{P(k_1k_2|D_i)P(D_i)}{\sum_{i=1}^{3} P(k_1k_2|D_i)P(D_i)} \tag{1}$$

We assume independence of symptoms k_1 and k_2, which results in

$$P(k_1k_2|D_1) = 0.2 \times 0.3 = 0.06$$

$$P(k_1k_2|D_2) = 0.4 \times 0.5 = 0.20$$

$$P(k_1k_2|D_3) = 0 \times 0.05 = 0$$

and the denominator in Eq (1) becomes

$$\sum_{i=1}^{3} P(k_1k_2|D_i)P(D_i) = 0.06 \times 0.05 + 0.20 \times 0.15 = 0.033$$

and

$$P(D_1|k_1k_2) = \frac{P(k_1k_2|D_1)P(D_1)}{0.033} \frac{0.06 \times 0.05}{0.033} = 0.09$$

$$P(D_2|k_1k_2) = \frac{P(k_1k_2|D_2)P(D_2)}{0.033} = \frac{0.20 \times 0.15}{0.033} = 0.91$$

$$P(D_3|k_1k_2) = \frac{P(k_1k_2|D_3)}{0.033} = \frac{0}{0.033} = 0$$

In complete analogy to the above for the second row

$$P(\bar{k}_1k_2|D_1) = P(\bar{k}_1|D_1)\ P(k_2|D_1) = [1 - P(k_1|D_1]P(k_2|D_1)$$

$$= 0.8 \times 0.3 = 0.24$$

$$P(\bar{k}_1k_2|D_2) = 0.6 \times 0.5 = 0.30$$

$$P(\bar{k}_1k_2|D_3) = 1 \times 0.05 = 0.05$$

so that

$$\sum_{i=1}^{3} P(k_1k_2|D_i)P(D_i) = 0.24 \times 0.05 + 0.30 \times 0.15 + 0.05 \times 0.8$$

$$= 0.097$$

and

$$P(D_1|\bar{k}_1k_2) = \frac{P(\bar{k}_1k_2|D_1)P(D_1)}{0.097} = \frac{0.24 \times 0.05}{0.097} = 0.124$$

$$P(D_2|\bar{k}_1k_2) = \frac{P(\bar{k}_1k_2|D_2)P(D_2)}{0.097} = \frac{0.30 \times 0.15}{0.097} = 0.464$$

$$P(D_3|\bar{k}_1k_2) = \frac{P(\bar{k}_1k_2|D_3)}{0.097}\frac{0.005 \times 0.80}{0.097} = 0.412$$

and for the third row

$$P(\bar{k}_1\bar{k}_2|D_1) = P(\bar{k}_1|D_1) = [1 - P(k_1|D_1)] \times [1 - P(k_2|D_1)]$$

$$= (1 - 0.2)(1 - 0.3) = 0.8 \times 0.7 = 0.42$$

$$P(\bar{k}_1\bar{k}_2|D_2) = 0.6 \times 0.5 = 0.30$$

$$P(\bar{k}_1\bar{k}_2|D_3) = 1 \times 0.95 = 0.95$$

Now

$$\sum_{i=1}^{3} P(\bar{k}_1\bar{k}_2|D_i)P(D_i) = 0.42 \times 0.05 + 0.30 \times 0.15 + 0.95 \times 0.8 = 0.82$$

and

$$P(D_1|\bar{k}_1\bar{k}_2) = \frac{P(\bar{k}_1\bar{k}_2|D_1)P(D_1)}{0.826} = \frac{0.42 \times 0.05}{0.826} = 0.0254$$

$$P(D_2|\bar{k}_1\bar{k}_2) = \frac{0.3 \times 0.15}{0.826} = 0.0545$$

$$P(D_3|\bar{k}_1\bar{k}_2) = \frac{0.95 \times 0.8}{0.826} = 0.9201$$

chapter 3

Single Random Variable

PROBLEM 3.1

A spring-mass system is subjected to harmonic sinusoidal excitation with specified frequency ω. Suppose the spring coefficient is a random variable with given probability distribution $F_K(K)$, where the mass m is a specified quantity. What is the probability of no resonance occurring in a system picked up at random?

SOLUTION 3.1

Denote: A-random event of the resonance occurring, the sought probability is then

$$P(\bar{A}) = 1 - P(A) = 1 - P(\Omega = \omega)$$

where $\Omega = \sqrt{\frac{K}{m}}$ or random variable, designating the random natural frequency. If K is a continuous random variable, then

$$P(\Omega = \omega) = P\left(\sqrt{\frac{K}{m}} = \omega\right) = P(K = m\omega^2)$$

i.e. K takes on a specific value $m\omega^2$, so that $P(\Omega = \omega) = 0$. Then $P(\bar{A}) = 1$.

The question can also be put in a different way: What is the probability of the natural frequency not exceeding the excitation frequency? This probability is

$$P\left(\sqrt{\frac{K}{m}} \leq \omega\right) = P(K \leq m\omega^2) = F_K(m\omega^2)$$

PROBLEM 3.2

A random variable X is said to have a *triangular distribution* if its density is

$$f_X(x) = \begin{cases} A\left(1 - \frac{|x|}{a}\right), & -a < x < a \\ 0, & \text{otherwise} \end{cases}$$

Find:

(a) the value of A,
(b) the mean value, $E(X)$
(c) the median, $med(X)$
(d) the *coefficient of variation*, $\gamma_X = \sigma_X / E(X)$
(e) the 0.9^{th} quantile, $\xi_{0.9}$.

SOLUTION 3.2

(a): Due to the "normalization" condition, given by Eq. (3.11):

$$\int_{-\infty}^{\infty} f_X(x) = 1$$

In our case

$$\int_{-a}^{a} A\left(1 - \frac{|x|}{a}\right) dx = 2\int_{0}^{a} A\left(1 - \frac{x}{a}\right) dx = 2\left(Ax - A\frac{x^2}{2a}\right)\Bigg]_0^a$$

$$= 2Aa - A\frac{a^2}{a} = Aa = 1$$

Hence $A = 1/a$

$$(b): E(X) = \int_{-a}^{a} x f_X(x) dx = \int_{-a}^{a} xA\left(1 - \frac{|x|}{a}\right) dx$$

$$= \int_{-a}^{0} xA\left(1 + \frac{x}{a}\right) dx + \int_{0}^{a} xA\left(1 - \frac{x}{a}\right) dx$$

$$= A\left\{\left(\frac{x^2}{2} + \frac{x^3}{3a}\right)\Bigg]_a^0 + \left(\frac{x^2}{2} - \frac{x^3}{3a}\right)\Bigg]_0^a\right.$$

$$= \frac{1}{a}\left\{-\frac{a^2}{2} + \frac{a^3}{3a} + \frac{a^2}{2} - \frac{a^3}{3a}\right\} = 0$$

The fact that the integral must vanish is seen immediately, since the limits of integration are symmetric about zero, $f_X(x)$ is an even function of x, and x is an odd function.

$$E(x^2) = a^2/6;\ \text{Var}(X) = a^2/6,\ 6_x = \frac{a}{\sqrt{6}}$$

(c): $F_X[\text{med}\ (X)] = 1/2$.

Since $f_X(x)$ is an even function of x,

$$\int_{-a}^{0} f_X(x)\,dx = \int_0^a f_X(x)dx$$

and therefore equal $1/2$.

Hence $F_X(0) = 1/2$

and med $(X) = 0$.

Another way: First calculate $F_X(x)$:

For $x \leq 0$:

$$F_X(x) = \int_{-a}^{x} A\left(1 + \frac{x}{a}\right)dx = \frac{1}{a}\left(x + \frac{x^2}{2a} + \frac{a}{2}\right)$$

For $0 < x \leq a$

$$F_X(x) = \int_{-a}^{0} A\left(1 + \frac{x}{a}\right)dx + \int_0^x A(1 - x/a)dx = \frac{1}{2} + \frac{1}{a}\left(x - \frac{x^2}{2a}\right)$$

For $x \leq -a$, $F_x(x) = 0$

For $x \geq a$, $F_x(x) = 1$

Now to find $y = \text{med}\ (x)$ we must solve the equation

$$F_X(y) = 1/2.$$

For $y \leq 0$, the latter equation becomes

$$\frac{1}{a}\left(y + \frac{y^2}{2a} + \frac{a}{2}\right) = \frac{1}{2}$$

which yields

$$y + \frac{y^2}{2a} = 0$$

The roots are

$$y_1 = 0,\ y_2 = -2a$$

but $F_X(-2a) = 0$, $F_X(0) = 1/2$, therefore med $(X) = 0$.

The same conclusion is reached for $y \geq 0$.

(d) To find the coefficient of variation, we first determine σ_X:

$$E(X^2) = \int_{-a}^{a} x^2 A\left(1 - \frac{|x|}{a}\right) dx = 2\int_0^a x^2 A\left(1 - \frac{x}{a}\right) dx$$

$$= 2A\left(\frac{x^3}{3} - \frac{x^4}{4a}\right)]_0^a = 2\frac{1}{a}\left(\frac{a^3}{3} - \frac{a^4}{4a}\right) = \frac{a^2}{6}$$

$$\gamma_X = \frac{\sigma_X}{E(X)} \to \infty$$

(e) $F_X(\zeta_{0.9}) = 0.9$

We first note that $F_X(0) = 1/2$, therefore $\zeta_{0.9} > 0$. Now $\frac{1}{2} + \frac{1}{a}\left(x - \frac{x^2}{2a}\right) = 0.9$ which leads to

$$x^2 - 2ax + 0.8a^2 = 0$$

$$x = a \pm 0.447a$$

$$\zeta_{0.9} = 0.553a.$$

This root has been chosen since $F_X(a + 0.447a) = F_X(1.447a) = 1 > 0.9$

PROBLEM 3.3

A random variable X is said to have a *beta distribution* if its density is

$$f_X(x) = \begin{cases} Ax^{\alpha-1}(1-x)^{\beta-1}, & 0 < x < 1 \\ 0, & \text{otherwise} \end{cases}$$

Find

(a) the value of A
(b) the distribution function associated with this probability density
(c) the probability of X taking on values less than 0.1
(d) the values of α and β for which a uniformly distributed random variables is obtained from X.

SOLUTION 3.3

The normalization condition (3.11) yields

$$A\int_0^1 x^{\alpha-1}(1-x)^{\beta-1} dx = 1$$

But

$$\int_0^1 x^{\alpha-1}(1-x)^{\beta-1}dx = \frac{\Gamma(\alpha)\Gamma(\beta)}{\Gamma(\alpha+\beta)} = B(\alpha,\beta)$$

and $A = 1/B(\alpha,\beta)$

Where $\Gamma(\alpha)$ is a gamma function, and $B(\alpha\beta)$ is known as a beta function-hence the name of the distribution, which reduces to the uniform distribution or the interval [0, 1] for $\alpha = \beta = 1$. We will show that

$$E(X) = \frac{\alpha}{\alpha+\beta}, \quad \text{Var}(X) = \frac{\alpha\beta}{(\alpha+\beta+1)(\alpha+\beta)^2}$$

Indeed,

$$E(X^k) = \frac{1}{B(\alpha,\beta)} \int_0^1 x^{k+\alpha-1}(1-x)^{\beta-1}dx$$

$$= \frac{B(k+\alpha,\beta)}{B(\alpha,\beta)} = \frac{\Gamma(k+\alpha)\Gamma(\beta)}{\Gamma(k+\alpha+\beta)} \cdot \frac{\Gamma(\alpha+\beta)}{\Gamma(\alpha)\Gamma(\beta)}$$

$$= \frac{\Gamma(k+\alpha)\Gamma(\alpha+\beta)}{\Gamma(\alpha)\Gamma(k+\alpha+\beta)}$$

Hence, for $k = 1$ we have

$$E(X) = \frac{\Gamma(\alpha+1)\Gamma(\alpha+\beta)}{\Gamma(\alpha)\Gamma(\alpha+\beta+1)} = \frac{\alpha}{\alpha+\beta}$$

and

$$\text{Var}(X) = E(x^2) - [E(X)]^2 = \frac{\Gamma(\alpha+\beta)}{\Gamma(\alpha)\Gamma(\alpha+\beta+2)} - \left(\frac{\alpha}{\alpha+\beta}\right)^2$$

$$+ \frac{(\alpha\delta 1)\alpha}{(\alpha+\beta+1)(\alpha+\beta)} - \left(\frac{\alpha}{\alpha+\beta}\right)^2 = \frac{\alpha\beta}{(\alpha+\beta+1)(\alpha+\beta)^2}$$

(b) The probability distribution function associated with the beta distribution is

$$F_X(x) = 0 \quad x < 0$$

$$= \frac{\Gamma(\alpha+\beta)}{\Gamma(\alpha)\Gamma(\beta)} \int_0^x u^{\alpha-1}(1-u)^{\beta-1}du, \; 0 \le x \le 1$$

$$1, \; x > 1$$

which can be integrated directly (say, numerically). For integer values of α and β direct analytical calculation is possible. For example, if $\alpha = \beta = 2$, then $0 \leq x \leq 1$

$$F_X(x) = \frac{\Gamma(4)}{\Gamma(2)\Gamma(2)} \int_0^x u(1-u)du = \frac{3!}{1!1!} \int_0^x (u - u^2)du$$

$$= 6\left(\frac{x^2}{2} - \frac{x^3}{3}\right)$$

For $\alpha = 2,\ \beta = 3$ we find directly

$$F_X(x) = \frac{\Gamma(5)}{\Gamma(2)\Gamma(2)} \int_0^x u(1-u)^2 du = \frac{4!}{1!2!} \int_0^x u(1 - 2u + u^2)du$$

$$= 12\left(\frac{x^2}{2} - \frac{2}{3}x^3 + \frac{x^4}{4}\right)$$

and so on.

In the general case the probability distribution function of a beta-variable is connected with the so-called incomplete beta function. The latter is usually denoted by $I_x(p, q)$:

$$F_X(x) \equiv F(x; \alpha, \beta) = \begin{cases} I_x(\alpha, \beta) & \text{if } \alpha \geq \beta \\ 1 - I_{(1-x)}(\beta, \alpha) & \text{if } \alpha < \beta \end{cases}$$

Note: In the context of Section 4.3 (Binomial or Bernoulli distribution) students can be asked to establish similarity between the probability densities of beta- and Bernoulli-distributed variables for α and β both positive integers. As a function of m, n and p, the probability of taking on the value $\gamma = m$ is for a binomially distributed variable γ (see Eq. 4.3)

$$P(\gamma = m) = \binom{n}{m} p^m q^{n-m} = \frac{n!}{m!(n-m)!} p^m q^{n-m}$$

Which, contrasted with the beta density

$$f_X(x) = \frac{(\alpha + \beta - 1)!}{(\alpha - 1)!(\beta - 1)!} x^{\alpha - 1}(1 - a)^{\beta - 1}$$

immediately establishes the relationship

$$f_X\,(x) = (\alpha + \beta - 1)\ P(\gamma = m)$$

if $P(\gamma = m)$ is evaluated at

$$m = \alpha - 1,\ n = \alpha + \beta - 2,\ p = x$$

For example, the value of $fx(0.7)$ with $\alpha = 2,\ \beta = 9$ is numerically equal to $7P(\gamma = 1)$ with $m = 1,\ n = 6$ and $P = 0.7$.

$$fx(0.5) = 7P(\gamma = 1) = 7\binom{6}{1}0.7^1 0.3^5 = 42 \times 0.7 \times 0.3^5$$

$$= 0.071492$$

Similarly, the relationship between $F_X(x)$ and $F_\gamma(m)$ is

$$f_x(x) = 1 - F_\gamma(m),\ \alpha\beta = 1, 2, \ldots x \le 1$$

with again $m = \alpha - 1,\ n = \alpha + \beta - 2,\ p = x$.

Say $\alpha = 9,\ \beta = 2$ and $x = 0.1$, then,

$$f_x = \begin{cases} \dfrac{\Gamma(11)}{\Gamma(9)\Gamma(2)} & x^8(1-x),\ 0 \le x \le 1 \\ 0, & \text{otherwise} \end{cases}$$

$F_X(0.1)$ is given by

$$F_X(0.1) = 1 - F_\gamma(k)$$

where γ is a binomial random variable with $k = \alpha - 1 = 8,\ n = \alpha + \beta - 2 = 9,\ p = 0.1$,

$$F_X(0.1) = 1 - F_\gamma(8) = 1 - \sum_{m=0}^{k \le 8} \binom{n}{m} 0.1^m \cdot 0.9^{9-m}$$

but $F_\gamma(8)$ equals the probability of the number of successes not exceeding 8 and since

$$\sum_{m=0}^{9} P(\gamma = m) = 1,$$

we have

$$F_\gamma(8) = 1 - P(\gamma = 9) + 1 - \binom{9}{9} 0.1^9 0.9^0 = 1 - 0.1^9$$

therefore,

$$F_X(0.1) = 1 - 1 + 0.1^9 = 0.1^9$$

(d) $\alpha = \beta = 1$, the uniformly distributed random variable in obtainable form X.

Remark: for $\alpha = 1,\ \beta = 2,\ F_X(x)$ becomes

$$F_X(x) = \begin{cases} 2(1-x), & \text{for } 0 \le x \le 1 \\ 0, & \text{otherwise} \end{cases}$$

and for $\alpha = 2$, and $\beta = 1$, $F_X(x)$ is

$$F_X(x) = \begin{cases} 2x, & \text{for } 0 \le x \le 1 \\ 0, & \text{otherwise} \end{cases}$$

These two distributions are known as triangular distributions.

Note to the lecturer: It is recommended to specify α and β as some positive integers (say $\alpha = 2,\ \beta = 3$) before presenting this problem, and ask the student to try the general case. They can be advised to find the above similarities between beta and binomial random variables, having first read and mastered Section 4.3.

PROBLEM 3.4

The *hazard function* $h(t)$ is defined as the conditional instantaneous failure rate. That is, $h(t)\,dt$ is the probability of the system failing in the time interval $(t,\ t+dt)$, given that the system has not failed prior to time t. It may be interpreted as the rate at which the system population still under test at time t is failing. Let T denote the random time of failure with density $f_T(t)$, such that $f_T(t)\,dt$ represents the probability of the system failing in the interval $(t,\ t+dt)$.

(a) Show that $h(t) = f_T(t|T \ge t)$

(b) Show that the hazard function at time t equals the probability density of the time of failure divided by the reliability $R(t)$, the probability of the system surviving up to t:

$$h(t) = \frac{f_T(t)}{R(t)},\ f_T(t) = F_T'(t),\ R(t) = 1 - F_T(t)$$

(c) Verify that

$$R(t) = [1 - F_T(0)] \exp\left[-\int_0^t h(t)dt\right]$$

$$f_T(t) = [1 - F_T(0)] h(t) \exp\left[-\int_0^t h(t)dt\right]$$

where $F_T(0)$ is the probability of failure at $t = 0$.

Remark: Since $0 \leq R \leq 1$, $h(t) \geq f_T(t)$. By analogy, the probability of a specimen subjected to a fatigue test with sufficiently high amplitude of the repeated load fracturing between 10^9 cycles [corresponding to $f_T(t)\,dt$] is very small. The probability of fracture in the same interval, provided the specimen survived up to 10^9 cycles [corresponding to $h(t)\,dt$] is much higher.

SOLUTION 3.4

(a) If $f_T(t)$ is an (absolute) probability density function one of the random time to failure, then $f_T(t)dt$ represents the probability of the system falling in the interval $(t, t+dt)$ [or the proportion of the population of systems starting at $t = 0$, which fail in that interval. On the other hand, $h(t)dt$ is the probability of a system not failing prior to t, but failing in the interval $(t, t+dt)$:

$$h(t)dt = P(t \leq T \leq t + dt | T \geq t)$$

or

$$h(t) = \frac{P(t \leq T \leq t + dt | T > t)}{dt} = f_T(t | T \geq t)$$

That is, $h(t)$ represents the condition probability density of a t-year-old system failing.

(b) let us show that

$$h(t) = \frac{f_T(t)}{R(t)}, \quad R(t) = 1 - F_T(t)$$

Indeed, according to definition of $h(t)$:

$$h(t)dt \equiv P(t \leq T \leq t + dt | T \geq t)$$

according to the definition of the conditional probability notion.
But,

$$P[t \leq T \leq t + dt) \cap (T \geq t)] = P[t \leq T \leq t + dt] \equiv f(t)dt$$

However,

$$P(t < T < t + dt | T > t) = \frac{P[t \leq T \leq t + dt) \cap)]}{P(T \geq t)}$$

according to the definition of the conditional probability notion.
But,

$$P[(t \leq T \leq t + dt) \cap (T \geq t)] = P[t \leq T \leq t + dt] \equiv f(t)dt$$

Since the event $(t \leq T \leq t + dt)$ is contained in $(T \geq t)$. On the other hand $P(T > t) = R(t)$

Where $R(t)$ is the system reliability-the probability of the system not failing prior to t (or of the time to failure exceeding t). Thus

$$h(t)dt = \frac{f_T(t)dt}{R(t)} \tag{3.1}$$

or

$$h(t)dt = \frac{f_T(t)}{R(t)} = \frac{f_T(t)}{1 - F_T(t)}$$

(c) Now

$$f_T(t) = \frac{dF_T(t)}{dt}$$

and with Eq. (3.1)

$$h(t)dt = \frac{dF_T}{1 - F_T}$$

integration yields

$$[-\ln(1 - F)]_{F_T(0)}^{F_T(t)} = \int_0^t h(t)dt$$

where $F_T(0)$ is the probability of initial failure at time $t = 0+$. Therefore,

$$ln\frac{1 - F_T(t)}{1 - F_T(0)} = -\int_0^t h(u)du$$

and

$$R(t) = 1 - F_T(t) = [1 - F_T(0)]\exp[-\int_0^t h(u)du]$$

or

$$R(t) = R(0)\exp[-\int_0^t h(u)du]$$

$$R(t) = R(0)\exp[-\int_0^t h(u)du]$$

the appropriate probability density is obtained by differentiation of Eq. (2)

$$-f_T(t) = -[1 - F_T(0)]h(t)\exp[-\int_0^t h(u)du]$$

PROBLEM 3.5

Find the probability density of the time failure $f_T(t)$, if the hazard function is constant $h(t) = a$, and the probability of initial failure is zero, $F_T(0) = 0$. Show that a is the reciprocal of the mean time of failure, $E(T)$.

SOLUTION 3.5

We apply the equation derived in Problem 3.4

$$-f_T(t) = -[1 - F_T(0)]h(t)\exp[-\int_0^t h(u)du]$$

for a constant hazard function, $h(t) = a$, and zero initial unreliability $F_T(0) = 0$, we have

$$f_T(t) = \alpha e^{-\int_0^t adu} = \alpha e^{-at}$$

and the mean life-time is

$$E(T) = \int_0^\infty t f_T(t)dt = \int_0^\infty tae^{-at}dt = \frac{e^{-at}}{a}(-at - 1)]_0^\infty = \frac{1}{a}$$

i.e. "a" is the reciprocal of the mean life-time.

PROBLEM 3.6

Find the conditional probability of a system failing in the time interval (t_1, t_2), assuming that it did not fail prior to time t_1; $h(t) = a$.

SOLUTION 3.6

$$P(t_1 \leq T \leq t_2 \,|\, T > t_1) = \frac{P[t_1 \leq T \leq t_2) \cap (T > t_1)]}{P(T > t_1)}$$

$$= \frac{P(t_1 \leq T \leq t_2)}{P(T > t_1)} = \frac{F_T(t_2) - F_T(t_1)}{1 - F_T(t_1)}$$

On the other hand, for $h(t) = a$, we have from Problem 3.5

$$f_T(t) = ae^{-at}$$

Whence

$$F_T(t) = 1 - e^{-at}$$

and finally,

$$P(t_1 \leq T \leq t_2 \,|\, T \geq t_1) = \frac{(1 - e^{-at_2}) - (1 - e^{-at_1})}{1 - (1 - e^{-at_1})}$$

$$= \frac{e^{-at_1} - e^{-at_2}}{e^{-at_1}} = 1 - e^{-a(t_2 - t_1)}$$

PROBLEM 3.7

Find the conditional probability of a system surviving in the time interval (t_1, t_2) assuming that it survived up to time t_1 [probability of prolongation by an additional time interval $\Delta t = t_2 - t_1$]; $h(t) = a$.

SOLUTION 3.7

The sought probability of prolongation is

$$P(T \geq t_2 \mid T \geq t_1) = \frac{P[(T \geq t_2) \cap (T \geq t_1)]}{P(T \geq t_1)}$$

Event $(T \geq t_2)$ is contained in $(T \geq t_1)$, therefore,

$$P[(T \geq t_2)|(T \geq t_1)] = P(T \geq t_1) = \frac{P(T \geq t_2)}{P(T > t_1)} = \frac{1 - F_T(t_2)}{1 - F_T(t_1)}$$

For $h(t) = \alpha$, we found in Problem 3.5

$$f_T(t) = ae^{-at}$$

The associated distribution probability is

$$F_T(t) = \int_0^t f_T(a)du = 1 - e^{-at}$$

and finally we have

$$P(T \geq t_2 \mid T \geq t_1) = \frac{1 - (1 - e^{-at_2})}{1 - (1 - e^{-at_1})} = \frac{e^{-at_2}}{e^{-at_1}} = e^{-a(t_2 - t_1)}$$

Denoting $t_2 - t_1 = \Delta t$, the last equation is rewritten as

$$P(T \geq t_1 + \Delta t \mid T \geq t_1) = e^{-a\Delta t}$$

this final answer can be interpreted as $P(T \geq \Delta t) = 1 - (1 - e^{-a\Delta t)} = e^{-a\Delta t}$or, in other words, as a statement that the probability of the system persisting at least $t + \Delta t$ hours given that it has survived t hours, in the same as the probability of it persisting at least Δt hours. Or, if the system is serviceable (alive) at time instant t_1, then the distribution of its remaining time of survival is the same as the original lifetime distribution, that is the system has no recollection of its already having been in use for a time t_1, it is said that the system is memoryless. An exponentially distributed random variable is memoryless in this sense.

PROBLEM 3.8

Verify that if X is uniformly distributed in the interval (a, b), the probability of $X \leq a + p(b - a)$, where $0 < p < 1$, equals p.

SOLUTION 3.8

For a uniformly distributed random variable we have (see page 54):

$$F_X(x) = \begin{cases} \dfrac{1}{b-a}, & \text{for } a \leq x \leq b \\ 0, & \text{otherwise} \end{cases}$$

and

$$F_X(x) = \begin{cases} 0, & \text{for } x \leq a \\ \dfrac{x-a}{b-a}, & \text{for } a \leq x \leq b \\ 1, & \text{for } b \leq x \end{cases}$$

We are interested in

$$P[x \leq a + p(b - a)]$$

Since $0 < p < 1$

$$a < a + p\,(b - a) < b$$

Therefore

$$P\,[X \leq a + p\,(b - a)] = F_X\,[a + p\,(b - a)] = \frac{[a + p\,(b - a)] - a}{b - a} = p$$

PROBLEM 3.9

Following the steps used in the text, prove the inequalities of Bienaymé and Tchebycheff for a discrete random variable.

SOLUTION 3.9

Let Y be a discrete random variable taking on positive values $\{y_i\}$, $i = 1, 2, \ldots,$ with probabilities $p_i = P(Y = y_i)$.

Then,

$$E(Y) = \sum_{i=1}^{\infty} y_i P_i = \sum_{i:y_1<\alpha} y_i P_i + \sum_{i:y_i\geq\alpha} y_i P_i$$

$$\geq \sum_{i:y_1\geq\alpha} y_i P_i \geq \alpha \sum_{i:y_i\geq\alpha} P_i = \alpha P(Y \geq \alpha)$$

Then the sought equation

$$P(Y \geq \alpha) \leq \frac{E(Y)}{\alpha}$$

follows immediately.

PROBLEM 3.10

X is uniformly distributed in the interval $[8, 12]$. Calculate the probability $P\{E(X) - \sigma_X < X < E(X) + \sigma_X\}$ and compare it with the upper bound furnished by Tchebycheff's inequality.

SOLUTION 3.10

For a uniformly distributed random variable (see page 54)

$$f_X(x) = \begin{cases} \dfrac{1}{b-a}, & \text{for } a \leq x \leq b \\ 0, & \text{otherwise} \end{cases}$$

in our case $a = 8$, $b = 12$, and

$$f_X(x) = \begin{cases} \dfrac{1}{4}, & \text{for } 8 \leq x \leq 12 \\ 0, & \text{otherwise} \end{cases}$$

We are interested in

$$P\{E(X) - \sigma_X < X < E(X) + \sigma_X).$$

Calculations yield (see page 55)

$$E(X) = \frac{a+b}{2} = \frac{8+12}{2} = 2$$

$$\text{Var}\,(X) = \frac{(b-a)^2}{12} = \frac{(12-8)^2}{12} = \frac{4}{3}$$

or by straightforward calculations

$$E(X) = \int_{-\infty}^{\infty} x f_X(x)dx = \int_{8}^{12} x \frac{1}{4}dx = \frac{x^2}{8}|_8^{12} = \frac{1}{8}\left(12^2 - 8^2\right) = 10$$

$$\text{Var}(X) = E(X^2) - [E(X)]^2 = \int_{-\infty}^{\infty} x^2 f_X(x)dx - 10^2 =$$

$$= \int_{8}^{12} x^2\frac{1}{4}dx - 10^2 = \left.\frac{x^3}{12}\right]_8^{12} - 10^2 = \frac{12^3 - 8^3}{12} - 100 = \frac{4}{3}$$

According to Tchebycheff's inequality (Eq. 3.58, p. 64) we have

$$P(|X - E(X)| \geq k\sigma_x) \leq \frac{1}{k^2}$$

Now for $k = 1$

$$P(|X - 10| \geq 1.1547) \leq 1$$

and the

$$P(|X - 10| \leq 1.1547 \geq 0$$

In this case Tchebycheff's inequality does not supply us with any new information, latter equations indicating non-negativeness of the sought probability.

PROBLEM 3.11

Show, using Tchebycheff's inequality, that if $E[(X - a)^2] = 0$, where a is a deterministic constant, then $X = a$ with unity probability.

SOLUTION 3.11

Direct proof of the statement is as follows:

The problem states that of $E[(X - a)^2] = 0$, then $X = a$ with unity probability, or that

$$P\{X \neq a\} = 0 \tag{3.2}$$

In fact, assume that the latter equation does not hold. We can find then an $\varepsilon > 0$ such that

$$P\{|X - a| > \varepsilon\} = \int |x - a| > \varepsilon\ f_X(x)dx \neq 0$$

On the other hand,

$$E[(X-a)^2] = \int_{-\infty}^{\infty} (x-a)^2 f_X(x)dx \geq \int |x-a| > \varepsilon\, (x-a)^2 f_X(x)dx$$

$$\geq \varepsilon^2 \int |x-a| > \varepsilon f(x)dx > 0$$

But this is impossible since it is given that $E[(X-a)^2]$ vanishes incidentally. Hence, Eq. (1) is valid.

Proof through Tchebycheff's inequality

According to Eq. (3.37)

$$P(|X-a| \geq \varepsilon) < \frac{E[|X-a|^n]}{\varepsilon^n}$$

We take here $n = 2$, yielding

$$P(|X-a| > \varepsilon) \geq \frac{E(X-a)^2}{\varepsilon^2}$$

But the right-hand side of this inequality is zero, according to the problem as given, i.e.,

$$P(|X-a| \geq \varepsilon \leq 0$$

This implies obviously, since a probability cannot be negative,

$$P(|X-a| \geq \varepsilon) = 0$$

or, that X takes on the value "a" with unity probability.

chapter 4

Examples of Probability Distribution and Density Functions. Functions of a Single Random Variable

PROBLEM 4.1

A dike is designed with 5 m freeboard above the mean sea level. The probability of its being topped by waves in 1 yr is 0.005. What is the probability of waves exceeding 5 m within 200 yrs?

SOLUTION 4.1

Let X be the number of large floods, caused by topping of a dike, over a period of 200 years, it is thus a binomially distributed random variable with $n = 200$ and $p = 0.005$. Hence

$$P(X > 1)1 - \binom{200}{0}(0.005)^0(1 - 0.005)^{200}$$

$$= 1 - (0.005)^{200}$$

Now

$$(1 - 0.005)^{200} = \left(1 - \frac{1}{200}\right)^{200} \simeq e^{-1}$$

and

$$P(X > 1)1 - e^{-1} = 1 - 0.368 = 0.632$$

Another question which may arise in connection with the dike problem is as follows. What is the probability of large flood occurring in the i-th year? A large flood in the i-th year is the first if and only if no such flood has taken place during the preceding $(i - 1)$ years; of which the probability is $(1 - 0.005)^i$, and such a flood occurs in the i-th year, of which the probability is 0.005. That is denoting by I the number of years to the first large flood,

$$P(I = 1) = (1 - 0.005)^i 0.005, \ i = 1, 2, 3, \ldots$$

In general, if probability of the dike being topped within 1 yr. is p,

$$P(I = i) = (1 - p)^i p$$

The distribution function is

$$\begin{aligned} F_I(i) &= \sum_{j=1}^{i} P(I = j) = \sum_{j=1}^{i} (1 - p)^j p \\ &= p \cdot \frac{(1 - p)^{i-1}}{(1 - p) - 1} = 1 - (1 - p)^i, \ i = 1, 2 \ldots \end{aligned}$$

It is worth noting, that this distribution is known as a geometric distribution with parameter p.

We are now in position to answer the following question: What is the probability of 20 years passing before a large flood occurs? I is a geometric random variable with parameter $p = 0.005$. The probability of $I > 20$ is

$$\begin{aligned} P(I > 20) &= 1 - F_I(20) = 1 - \{1 - (1 - 0.005)^{20}\} \\ &= (1 - 0.005)^{20} \end{aligned}$$

PROBLEM 4.2

Using Eq. (4.23) for the characteristic function of a normally distributed random variable, show that

$$E(X) = a \quad \mathrm{Var}(X) = \sigma_X^2$$

SOLUTION 4.2

$$M_X(\Theta) = \exp\left(i\Theta a - \frac{1}{2}\sigma_X \Theta^2\right)$$

$$E(X) = \frac{-1}{1} \left.\frac{dM_X(\Theta)}{d\Theta}\right|_{\Theta=0} = \frac{-1}{1}\left.\left(ia - \frac{1}{2}\sigma_X^2\Theta^2\right)\right|_{\Theta=0}$$

$$= a$$

For $\mathrm{Var}(X)$ we have

$$\mathrm{Var}(X) = m_2 - m_1^2$$

Now

$$m_2 = \frac{1}{i^2}\left.\frac{d^2 M_X(\Theta)}{d\Theta^2}\right|_{\Theta=0} = -\left\{(ia - \sigma_X^2\Theta)^2 e^{i\Theta a} - \frac{1}{2}\sigma_x^{2\Theta^2}\right.$$

$$\left.\left. -\sigma_X^2 e^{i\Theta a - \frac{1}{2}\sigma_X^2\Theta^2}\right\}\right|_{\Theta=0} a^2 + \sigma_X^2$$

and

$$\mathrm{Var}(X) = a^2 + \sigma_X^2 - a^2 = \sigma_X^2$$

Another way of obtaining the variance is via Eq. 3-48

$$\mathrm{Var}(X) = -\left|\frac{d^2 lnm_x(\theta)}{d\theta^2}\right| \theta = 0$$

$$= -\left|\frac{d^2}{d\theta^2}\left(i\theta a - \frac{1}{2}\sigma_x^2\theta^2\right)\right|_{\theta=0} = \sigma_x^2.$$

PROBLEM 4.3

Suppose that the duration of successful performance (lifetime, in years) of a piece equipment is normally distributed with a mean of 8 yrs. What is the largest value its standard deviation may have if the operator requires at least 95% of the population to have lifetimes exceeding 6 yrs? What is the probability of a piece of equipment turning out faulty at delivery?

SOLUTION 4.3

We are interested in the largest value of the standard deviation σ_T of the successful performance life (in years) T of a piece of equipment:

$$P(T > 6) \geq 0.95$$

$$P(T > 6) = 1 - P(T \leq 6) = 1 - \left[\frac{1}{2} + \text{erf}\left(\frac{6-8}{\sigma_T}\right)\right]$$

$$= \frac{1}{2} - \text{erf}\left(-\frac{2}{\sigma_T}\right) = \frac{1}{2} + \text{erf}\left(\frac{2}{\sigma_T}\right) \geq 0.95$$

The maximum value of σ_T is found for

$$\frac{1}{2} + \text{erf}\left(\frac{2}{\sigma_T, max}\right) = 0.95$$

or

$$\text{erf}\left(\frac{2}{\sigma_T, \max}\right) = 0.45, \quad \frac{2}{\sigma_T, \max} = \text{erf}^{-1}(0.45)$$

We find by interpolation from the table of the error function on page 471, that

$$\text{erf}^{-1}(0.45) = 1.6448543$$

Hence

$$\sigma_{T,max} = \frac{2}{1.6448543} = 1.216 \text{ years}$$

The probability of a piece of equipment turning out faulty at delivery is

$$P(T \leq 0) = \frac{1}{2} + \text{erf}\left(\frac{0-8}{1.216}\right) = \frac{1}{2} - \text{erf}(6.5789) \simeq 0$$

PROBLEM 4.4

A system is activated at $t = 0$. Its time of failure is the random variable T with distribution function $F_T(t)$ and density $f_T(t)$. Denote the hazard function by $h(t)$ (see Prob. 3.4)

(a) Does T in Prob. 3.4 have an exponential distribution?
(b) Show that if $h(t) = \alpha t$, T has a Rayleigh distribution.
(c) Show that if $h(t) = \alpha\beta t^{\beta-1}$, T has a Weibull distribution.
(d) Find $f_T(t)$ if $h(t) = \alpha\gamma e^{\gamma t}$.

Remark The resulting $f_T(t)$ is called an *extreme-value* density function. For interesting failure models leading to the extreme-value distribution, see Epstein (1958).

Note to Lecturer: In question (a), read "prob. 3.5" for "prob. 3.4."

SOLUTION 4.4

(a) According to Prob. 3.5, we have:

$$f(t) = [1 - F_T(0)]h(t)\exp\left[-\int_0^t h(u)du\right.$$

For $h(t) = a$ and $F_T(0) = 0$, we found Prob. 3.5

$$f_T(t) = ae^{-at}$$

Which implies that T has an exponential distribution.

(b) $h(t) = \alpha t$, Then

$$f_T(t) = h(t)\exp\left[-\int_0^t h(u)du\right] = \alpha t\exp\left[-\int_0^t \alpha u\,du\right]$$

$$= \alpha te^{\frac{-\alpha t^2}{2}}$$

which implies that T has a Rayleigh distribution (see Eq. 4-1) with

$$\alpha = \frac{1}{a^2}$$

(c) $h(t) = \alpha\beta t^{\beta-1}$, then

$$f_T(t) = h(t)\exp\left[-\int_0^t h(u)du\right] = \alpha\beta t^{\beta-1}\exp\left[-\int_0^t \alpha\beta u^{\beta-1}du\right]$$

$$= \alpha\beta t^{\beta-1}e^{-\alpha t\beta}$$

which implies that T has a Weibull distribution (see Eq. 4–11).

(d) $h(t) = \alpha\gamma e^{\gamma t}$, then

$$f_T(t) = h(t)\exp\left[-\int_0^t h(u)du\right. = \alpha\gamma e^{\gamma t}\exp\left[-\int_0^t \alpha\gamma e^{\gamma u}du\right]$$

$$= \alpha\gamma e^{\gamma t e^{(\alpha e^{\gamma t}-\alpha)}} = \alpha\gamma e^{\alpha(1-e^{\gamma t})+\gamma t}$$

If we denote

$$a = \alpha e^{\gamma t}$$

then $f_T(t)$ becomes

$$f_T(t) = \alpha\gamma e^{\alpha-a}$$

PROBLEM 4.5

A system consisting of n elements in parallel with equal reliabilities $R(t)$ fails only when all elements fail simultaneously.

(a) Show that the reliability $R_n(t)$ of such a system is

$$R_n(t) = [1 - R(t)]^n$$

(b) Find $R_n(t)$ when the conditional failure rate of each element is constant, $h(t) = a$.

(c) Find the *allowable operation time* t, of the system, such that $R(t_r) = r$, where r is a required reliability, for $n = 1, 2, 3$.

(d) Show that the mean lifetime of the system is

$$E(T) = \frac{1}{a}\left(1 + \frac{1}{2} + \cdots + \frac{1}{n}\right)$$

and interpret the result for $n \to \infty$.

SOLUTION 4.5

(a) Consider first the unequal reliabilities.

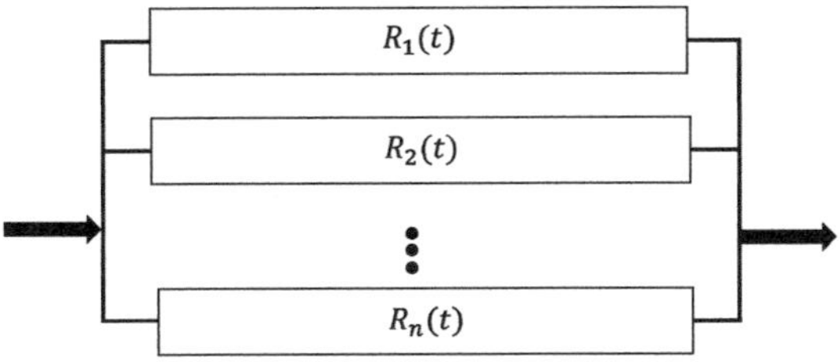

Since the elements are connected in parallel the system fails when all of them fail. The probability of failure is

$$P(A_1A_2\ldots A_n) = P(A_1)P(A_2)\ldots P(A_n)$$

$$= [1 - R_1(t)][1 - R_2(t)]\ldots[1 - R_n(t)] = \prod_{j=1}^{n}[1 - R_j(t)]$$

Where A_j denotes failure of the jth element. The reliability of the entire system is

$$R_n(t) = 1 - \sum_{j=1}^{n}[1 - R_j(t)]$$

For equal reliabilities $R_j(t) \equiv R(t)$ we obtain the sought result

$$R_n(t) = 1 - [1 - R(t)]^n$$

(b) If $h(t) = a$, then, by the formula for $R(t)$ in Prob. 3.4c,

$$R(t) = [1 - F_T(0)]\exp\left[-\int_0^t h(t)dt\right]$$
$$= [1 - F_T(0)]e^{-at}$$

where $F_T(0)$ is the probability of failure at $t = 0$. Therefore

$$1 - F_T(0) = R(0)$$

and $R(t) = R(0)e^{-at}$
For the reliability of the system we have

$$R_n(t) = 1 - [1 - R(0)e^{-at}]^n$$

If the probability of failure of $t = 0$ vanishes, $F_T(0) = 0$ and $R(0) = 1$,

$$R_n(t) = 1 - [1 - e^{-at}]^n$$

(c) The allowable (or design) operation time t_r is defined as the solution of the equation

$$R_n(t_r) = r$$

or

$$1 - [1 - e^{-at}r] = r$$

which yields

$$t_r = -\frac{1}{a}ln[1 - (1 - r)^{1/n}]$$

For $a = 0.001$ and $r = 0.999$ we obtain

$$n = 1: \quad t_r \simeq 1$$
$$n = 2: \quad t_r \simeq 32$$
$$n = 3: \quad t_r \simeq 105$$

Implying that the design time increases with the number of elements connected in parallel.

(d) By definition of the mathematical expectation,

$$E(T) = \int_{-\infty}^{\infty} t f_T^{(n)}(t)dt$$

or, since $T \geq 0$,

$$E(T) = \int_0^\infty t f_T^{(n)}(t)dt$$

where $f_T(n)(t)$ is the probability density of the failure time for the system of n elements in question.
But

$$f_T(t) = -\frac{dR_n(t)}{dt}$$

Therefore

$$E(T) = -\int_0^\infty t\frac{dR_n(t)}{dt}$$

Integration by parts yields,

$$E(T) = -tR_n(t)]_0^\infty + \int_0^\infty R_n(t)dt$$

or, for our specific problem, [with $R(0) = 1$]:

$$E(T) = -t[1 - (1 - e^{-at})^n]_0^\infty + \int_0^\infty [1 - (1 - e^{-at})^n]dt$$

However,

$$\lim_{t\to\infty} t[1 - (1 - e^{-at})^n] = 0$$

and

$$E(T) = \int_0^\infty [1 - (1 - e^{-at})^n]dt$$

We substitute

$$1 - e^{-at} = y$$

$$ae^{-at}dt = dy$$

With the former equation in mind

$$dt = \frac{1}{a}\frac{1}{1-y}dy$$

where

$$E(T) = \frac{1}{a}\int_0^1 (1-y)^n \frac{dy}{1-y}$$

since

$$t = 0, \; y = 0$$

$$t \to \infty, \; y = 1$$

Now

$$\frac{(1-y)^n}{1-y} = 1 + y + y^2 + \cdots y^{n-1}$$

and

$$E(T) = \frac{1}{a}\int_0^1 (1 + y + y^2 + \cdots + y^{n-1})dy$$

$$= \frac{1}{a}\left(1 + \frac{1}{2} + \frac{1}{3}\cdots + \frac{1}{n}\right), \quad Q.E.D.$$

for $n \to \infty$, we obtain a divergent, harmonic series

$$\sum_{j=1}^{\infty}\frac{1}{j} \to \infty$$

implying that the mean life time increases indefinitely in a parallel system.

PROBLEM 4.6

Repeat Problem 4.5, but with the elements arranged in series instead of in parallel.

SOLUTION 4.6

The reliability of the system is (see Eq. 2.17):

$$R_n(t) = [R(t)]^n$$

If the lifetime of each element is distributed exponentially, then

$$R_n(t) = [R(0)e^{-at}]^n = [R(0)]^n e^{ant}$$

For
$R(0) = 1$, we get

$$R_n(t) = e^{-ant}$$

The design operation time t_r of the system, such that $R_n(t_r) = r$ where r is the required reliability, is found as

$$e^{-ant_r} = r$$

which yields

$$t_r = \frac{1}{an} ln \frac{1}{r}$$

indicating decrease of t_r with increase of the number of elements n. The probability of failure of the system equals the distribution function of the time of failure

$$F_T(t) = 1 - e^{-ant}$$

implying that T is also distributed exponentially, with

$$E(T) = \frac{1}{an}$$

$E(T)$ tends to zero for n increasing indefinitely.

PROBLEM 4.7

Suppose that the reliabilities of n individual elements are identical and are given by

$$R(t) = t/\tau \quad \text{for} \quad 0 \le t \le \tau$$

Show that the mean lifetime of the system is

$$E(T) = \begin{cases} r/(n+1) \text{ for elements in series} \\ rn/(n+1) \text{ for elements in parallel} \end{cases}$$

Interpret the result for $n \to \infty$.

SOLUTION 4.7

Note for lecturer: Reliabilities of individual elements are $R(t) = 1 - t/\tau$ for $0 \le t \le r$, (see Eq. 2.17)

For a series system ($0 \le t \le \tau$) we have

$$R_n(t) = R^n(t) = \left(1 - \frac{t}{\tau}\right)^n$$

$$E(T) = \int_0^\tau R_n(t)dt = \int_0^\tau \left(1 - \frac{t}{\tau}\right)^n dt = \frac{\tau}{n+1}$$

For a parallel system (see Eq. 2.18)

$$R_n(t) = 1 - \left(1 - \frac{t}{\tau}\right)^n$$

$$E(T) = \int_0^\tau \left[1 - \left(\frac{t}{\tau}\right)^n\right] dt = \tau - \int_0^\tau \left(\frac{t}{\tau}\right)^n dt$$

which yields

$$E(T) = \tau - \frac{\tau^{h+1}}{(h+1)\tau^n} = \tau - \frac{\tau}{n+1} = \frac{n}{n+1}\tau$$

Note that for fixed τ, when the number of components becomes very large, $E(T)$ tends to zero, a series system and to τ for a parallel system. Recall that the mean lifetime of each component is $\tau/2$.

Indeed

$$E(T) = \int_0^\tau R(t)dt = \int_0^\tau \left(1 - \frac{t}{\tau}\right) dt = \tau - \frac{\tau^2}{2.\tau} = \frac{\tau}{2}.$$

PROBLEM 4.8

The random variables X and Y are linked the following functional relationship

$$Y = \begin{cases} -a^3, & x < -a \\ X^3, & -a \le x \le a \\ a^3, & x > a \end{cases}$$

X has a uniform distribution in the interval $(-a, a)$. Find $F_Y(y)$ and $f_Y(y)$, and present them graphically.

SOLUTION 4.8

$$f_X(x) = \begin{cases} \dfrac{1}{2a}, & \text{for } a- \le n \le a \\ 0, & \text{otherwise} \end{cases}$$

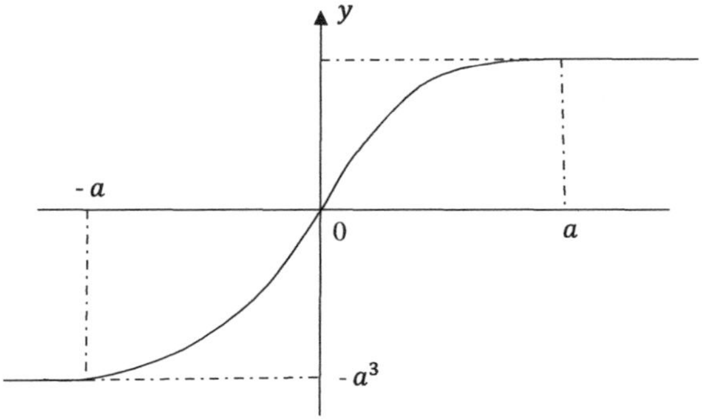

Since Y cannot take on values less than $-a^3$, $F_Y(y) = 0$ for $y \leq -a^3$. Since Y is always less than a^3, $F_Y(y) = 1$, for $y \leq a^3$. In the interval $-a^3 \leq y \leq a^3$, the inequality $Y \leq y$ is equivalent to

$$X^3 \leq y$$

and therefore

$$F_Y(y) = P(Y \leq y) = P(X \leq y^{1/3}) = F_X(y^{1/3})$$

But

$$F_X(x) = \begin{cases} 0, & \text{for } x \leq -a \\ \dfrac{x+a}{2a}, & \text{for } -a \leq x \leq a \\ 1, & \text{for } x \leq a \end{cases}$$

Hence

$$F_Y(y) = \begin{cases} 0, & \text{for } y^2 \leq a^3 \\ \dfrac{y^{1/3}+a}{2n}, & \text{for } -a^3 \leq y \leq a^3 \\ 1, & \text{for } y \leq a^3 \end{cases}$$

Now

$$fY(y) = \begin{cases} 0, & \text{for } y^2 - a^3 \\ \dfrac{1}{6a}y^{-2/3} & \text{for } -a^3 \leq y \leq a^3 \\ 0, & \text{for } y \leq a^3 \end{cases}$$

PROBLEM 4.9

Given the functions of Problem 4.8 but with having X having a uniform distribution in the interval $(-2a, 2a)$, show that Y is a mixed random variable.

SOLUTION 4.9

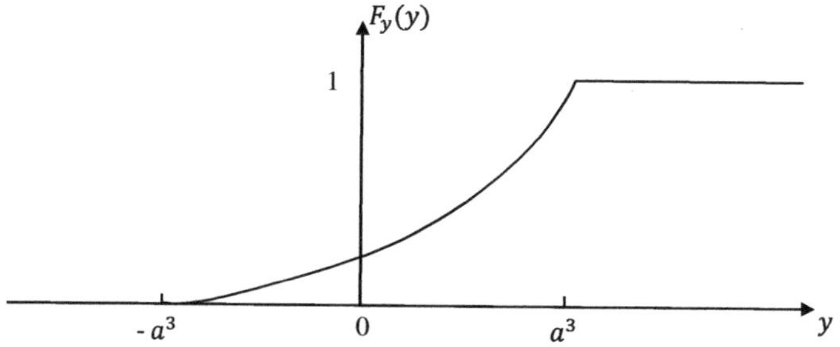

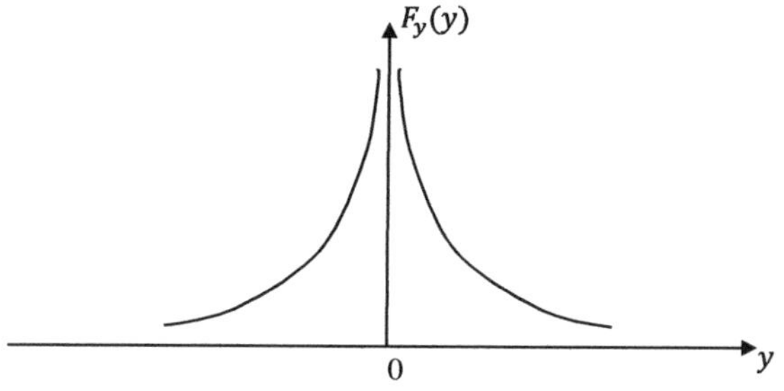

Now

$$f_X(x) = \begin{cases} \dfrac{1}{4a}, & \text{for } -2a \leq x \leq 2a \\ 0, & \text{otherwise} \end{cases}$$

$$F_X(x) = \begin{cases} 0, & \text{for } x \leq -2a \\ \dfrac{x+2a}{4a}, & \text{for } -2a \leq x \leq 2a \\ 1, & \text{for } x \leq 2a \end{cases}$$

and, as before

$$F_Y(y) = F_X(y^{1/3})$$

which yields

$$F_Y(y) = \begin{cases} 0, & \text{for } y \leq -a^3 \\ \dfrac{y^{1/3} + 2a}{4a}, & \text{for } -a^3 < y < a^3 \\ 1, & \text{for } y \geq a^3 \end{cases}$$

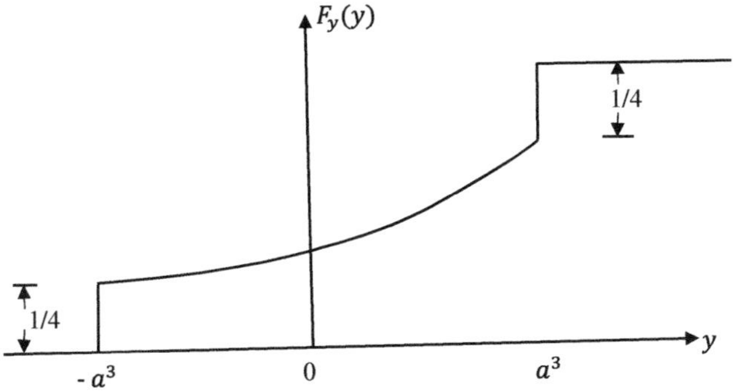

This function exhibits equal jumps at $y = -a^3$ and $y = a^3$. The jumps of the distribution functions, according to equations (3.16) and (3.17), represent a linear combination of delta functions with the coefficients equal the values of the jumps. Therefore, the probability density function is expressed as follows

$$f_Y(0) = \frac{1}{12a} y^{-\frac{2}{3}} [U(y + a^3) - U(y - a^3)]$$
$$+ \frac{1}{4}\delta(y + a^3) + \frac{1}{4}\delta(y - a^3)$$

Thus, the random variable Y has continuous set of possible values, filling the interval $(-a^3, a^3)$ and, in addition, two specific values $-a^3$ and a^3 with probability $1/4$, indicating that Y is a mixed random variable.

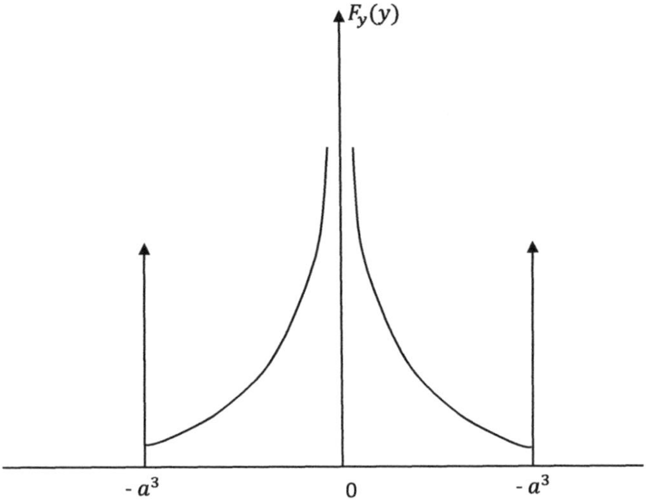

PROBLEM 4.10

X has a Cauchy distribution. Show that $Y = 1/X$ also has a Cauchy distribution.

SOLUTION 4.10

We use Eq. (4.45)

$$f_Y(y) = f_X[\psi(y)]\left|\frac{d\psi(y)}{dy}\right|, \quad f_X(x) = \frac{a}{\pi}\frac{1}{a^2+x^2}$$

for $Y = \phi(X)$, where $\phi(x)$ is a monotone function, as in our case

$$y = \psi(x) = \frac{1}{x}$$

and

$$x = \psi(y) = \frac{1}{y}.$$

Now

$$\left|\frac{d\psi(y)}{dy}\right| = \left|\frac{-1}{y^2}\right| = \frac{1}{y^2}$$

and

$$f_Y(y) = \frac{a}{\pi}\frac{1}{a^2 + (1/y^2)}\frac{1}{y^2} = \frac{b}{\pi}\frac{1}{y^2 + b^2}$$

implying that Y has a Cauchy density with parameter $b = 1/a$.

Another solution is as follows:

The probability distribution function of X is (see Example 3.8, pg. 51)

$$F_X(x) = \frac{1}{2} + \frac{1}{\pi}\tan^{-1}\left(\frac{x}{a}\right)$$

Now for $Y = 1/X$ we have

$$F_Y(y) = P(Y \leq y) = P\left(\frac{1}{X} \leq y\right)$$

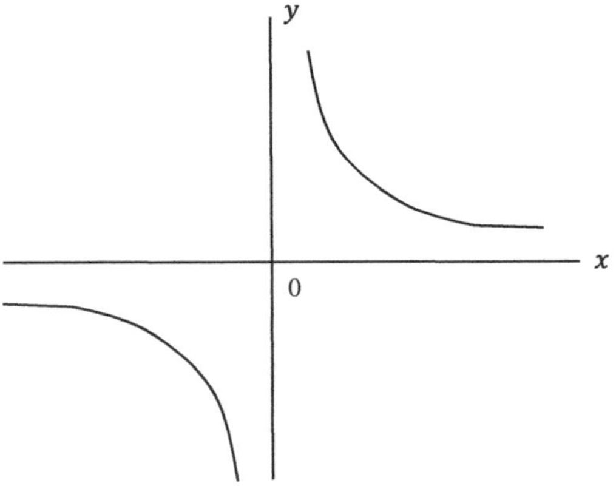

consider first the case $y < 0$

$$F_Y(y) = P\left(\frac{1}{y} \leq X \leq 0\right) = F_X(0) - F_X\left(\frac{1}{y}\right) = \frac{1}{2} - F_X\left(\frac{1}{y}\right)$$

and

$$f_Y(y) = \frac{dFy(y)}{dy} = \frac{1}{y^2}f_X\left(\frac{1}{y}\right) = \frac{1}{y^2}\frac{a}{\pi}\frac{1}{a^2 + (1/y)^2} = \frac{(1/a)}{\pi}\frac{1}{(1/a^2) + y^2}$$

$y < 0$ (1)

For the case $y > 0$, we have

$$F_Y(y) = P\left[(X < 0) U P\left(X < \frac{1}{y}\right)\right] = P(X < 0) + P\left(X > \frac{1}{y}\right)$$

$$= F_X(0) + 1 - F_X\left(\frac{1}{y}\right) = 1.5 - F_X\left(\frac{1}{y}\right)$$

For the probability density, we have

$$f_Y(y) = \frac{1}{y^2} f_X\left(\frac{1}{y}\right) = \frac{1}{y^2}\frac{a}{\pi}\frac{1}{a^2 + 1/y^2} \tag{2}$$

Equations (1) and (2) indicate that Y has a Cauchy distribution,

$$f_Y(y) = \frac{b}{\pi}\frac{1}{(1/b^2) + y^2}, \quad -\infty < y < \infty$$

where

$$b = 1/a.$$

chapter 5

Reliability of Structures Described by a Single Random Variable

PROBLEM 5.1

The loads acting on different machine components often have a chi-square distribution with m degrees of freedom, as was shown by Serensen and Bugloff,

$$f_N(n) = \frac{e^{-\frac{n}{2}} n^{r-1}}{2^{\nu}\Gamma(\nu)} U(n),\ \nu = \frac{m}{2}$$

Find the reliability of the bar under random tensile force. Table 5.1 contains the values of $\chi^2_{\alpha,\nu}$, where $\int_{\chi^2_{\alpha,\nu}}^{\infty} f_N(n)dn = \alpha$ for $\alpha = 0.995, 0.99, 0.975, 0.05, 0.025, 0.01, 0.005$, and $\nu = 1, 2, \ldots, 30$ (see figure).

SOLUTION 5.1

Notes to lecturer

Note 1. The chi-square distributed random variable X with

$$f_X(x) = \frac{e^{-\frac{x}{2}} x^{\nu-1}}{2^{\nu}\Gamma(\nu)} U(x), \nu = m/2$$

is nondimensional. For it to represent an axial force, we will use the transformation

$$N = n_0 X,\ n_0 > 0$$

where n_0 is a constant with the dimension of force. Then (see Equation 4.46)

$$f_N(n) = f_X\left(\frac{n}{n_0}\right)\frac{1}{n_0}$$

and

$$f_N(n) = \frac{e^{-\frac{n}{2n_0}} n^{\nu-1}}{2^\nu n_0^\nu \Gamma(\nu)}, \quad n > 0$$

If N is measured in say, kilonewtons, we can choose $n_0 = 1\,kN$, which yields the formula given in the problem.

Note 2. This problem can be designed in three forms; σ's the first, the second or the third basic problem of the probabilistic strength of materials.

First problem: Lecturer should specify the cross-sectional area of the bar "a", yield stress σ_Y and required reliability r, as well as "m" the number of degrees of freedom of x^2-distribution. The reliability is:

$$R = F_N(\sigma_Y a)$$

Table 5.1 lists the values of $x^2\alpha$, ν, so that

$$\int_{x^2}^{\infty} \alpha, \nu f_N(n)dn = \alpha$$

Now

$$R = F_N(\sigma_Y a) = \int_0^{\sigma_Y a} f_N(n)dn = 1 - \int_{\sigma_Y a}^{\infty} f_N(n)dn$$

Thus if $\sigma_Y a$ is chosen such that

$$\sigma_Y a = x^2_{\alpha,\nu}$$

we have

$$R = 1 - \alpha$$

if $1 - \alpha > r$, the strength requirement is met, and violated otherwise.

For the numerical values to be valid, the following set of data should be given to student (as an example):

$$\nu = 20, \quad \sigma_Y = 6\,KN/mm^2, \quad a = 5.695mm^2, \quad r = 0.95$$

Then $\sigma_Y a = 6 \cdot 5.695 = 34.17$,

$$R = F_N(\sigma_Y a) = 1 - \alpha = 1 - 0.025 = 0.975 > 0.95$$

which means that strength requirement is met.

Second problem: Determination of maximum admissible ν.

Numerical example:
Find max ν, such that $R \geq r = 0.99$, $\sigma_Y = 22kN/mm^2$, $a = 2mm^2$.
Solution: $\alpha = 1 - R \leq 0.01$, $\sigma_Y a = 44kN$, from Table 5.1 we find $\nu_{max} = 25$.

Third Problem: Design.

Find "a", so that $R \geq r = 0.995$, $\sigma_Y = 22kN/mm^2$, $\nu = 16$. From Table 5.1 for $\alpha = 1 - R = 1 - 0.995 = 0.005$, we find for $\nu = 16, x^2\alpha, \nu = 34.267 = \sigma_Y a$;

$$a_{req} = \frac{34.267}{22} = 1.56\,mm^2$$

PROBLEM 5.2

Find the reliabilities of the truss structures shown in Figs. 2.9 a, c, d and e, if P is treated as a random variable with given density function $f_P(p)$.

SOLUTION 5.2

Note to lecturer: the lengths of all bars in figure 2.9.a. are equal (denote l).

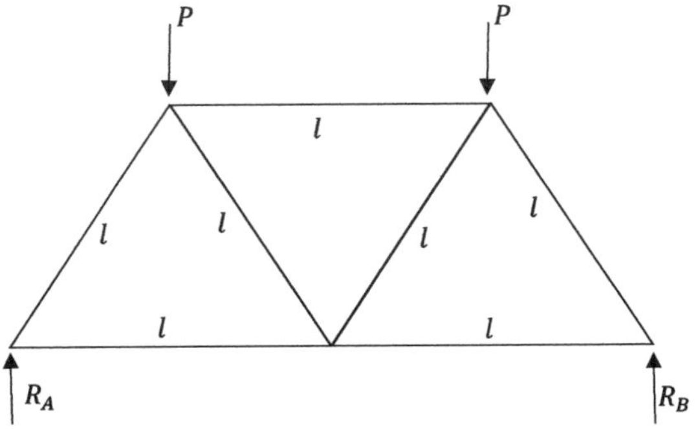

$$\frac{\sum M_A = 0}{R_B \cdot 2l} - P\frac{a}{2} - 2P\tfrac{3l}{2} = 0$$

$$R_B = \frac{7}{4}P \qquad \frac{\sum M_B = 0}{P \cdot \frac{3l}{2} + 2P\frac{l}{2} - R_A \cdot 2l = 0} \qquad R_A = \frac{5}{4}P$$

Check: $\sum Y = 0: \; R_A + R_B = 3P; \; \frac{7}{4}P + \frac{5}{4}P = 3P$

We use the method of sections to find the axial loads in each of the bars:

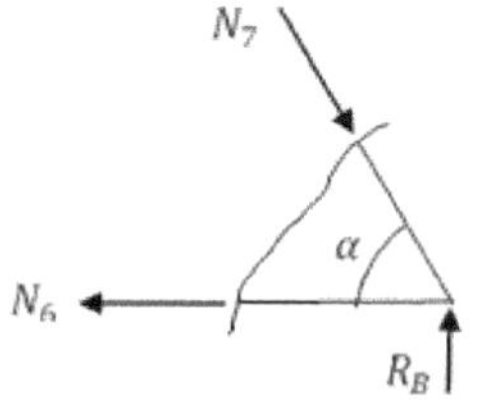

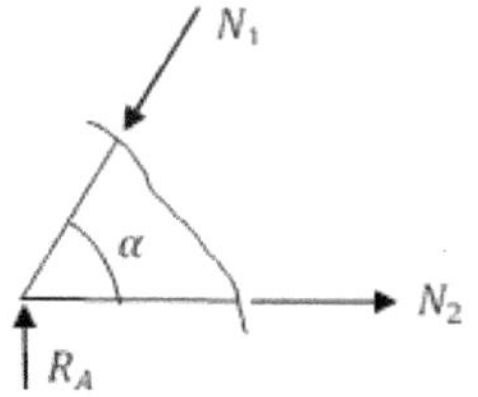

$$\sum Y = 0 - N_1 \sin\alpha + R_A = 0 \quad N_1 = \frac{R_A}{\sin 60^\circ} = \frac{5}{4}\frac{P}{\sqrt{3/2}} = \frac{5}{2\sqrt{3}}P$$

$$\sum X = 0 N_2 - N_1 \cos\alpha = 0 \quad N_2 = \frac{1}{2}N_1 = \frac{5}{4\sqrt{3}}P$$

$$\sum Y = 0 - N_7 \sin\alpha + R_B = 0 \quad N_7 = \frac{R_B}{\sin 60^\circ} = \frac{7}{4}\frac{P}{\sqrt{3/2}} = \frac{7}{2\sqrt{3}}P$$

$$\sum^{\leftarrow} X = 0 N_6 - N_7 \cos\alpha = 0, \quad N_6 = \frac{1}{2}N_7 = \frac{7}{4\sqrt{3}}$$

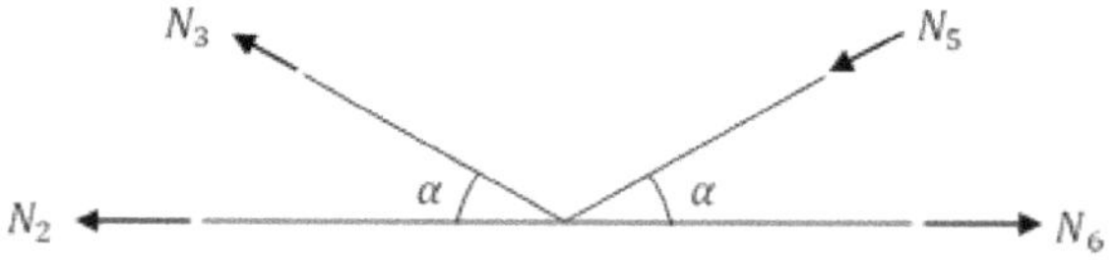

$$\uparrow \sum Y = 0 \quad N_3 \sin\alpha - N_5 \sin\alpha = 0$$

$$N_3 = N_5$$

$$\sum X = 0 \quad N_6 - N_2 - N_3 \cos\alpha = N_5 \cos\alpha = 0 \; N_6 + N_2 + 2N_3 \cos\alpha$$

$$= N_2 + 2 \cdot N_3 \cdot \frac{1}{2} = N_2 + N_3$$

Therefore

$$N_3 = N_5 = N_6 - N_2 = P\left(\frac{7}{4\sqrt{3}} - \frac{5}{4\sqrt{3}}\right) = \frac{P}{2\sqrt{3}}$$

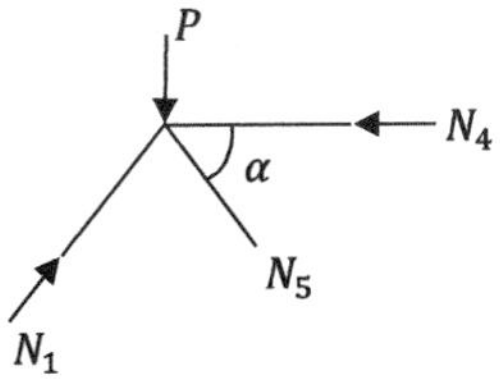

$$\downarrow \sum Y = 0; \quad P + N_3 \sin\alpha - N_1 \sin\alpha = 0$$

since we already know N_1 and N_3 this equation serves as a check.
Indeed

$$P + \frac{P}{2\sqrt{3}} \cdot \frac{\sqrt{3}}{2} - \frac{5}{2\sqrt{3}}P \cdot \frac{\sqrt{3}}{2}P \equiv 0$$

$$\sum^{\leftarrow} X = 0; N_4 = (N_1 + N_3)\cos\alpha = P\left(\frac{5}{2\sqrt{3}} + \frac{1}{2\sqrt{3}}\right) \cdot \frac{1}{2}$$

$$= \frac{3}{2\sqrt{3}}P = \frac{\sqrt{3}}{2}P$$

Finally:

$$N_1 = \frac{5}{2\sqrt{3}}P = 1.4434\,P \quad N_2 = \frac{5}{2\sqrt{3}}P = 0.72169\,P$$

$$N_3 = \frac{P}{2\sqrt{3}} = 0.288675P \quad N_4 = \frac{\sqrt{3}}{2}P = 0.866\,P$$

$$N_5 = N_3 = 0.288675\,P \quad N_6 = \frac{7}{4\sqrt{3}}P = 1.0104\,P$$

$$N_7 = \frac{7}{2\sqrt{3}}P = 2.0207\,P$$

As we see, the maximum load acts in the seventh bar. Its failure implies failure of the entire stress. In other words:

$$R = R_7 = \text{Prob}\left(\left|\frac{7}{2\sqrt{3A}}\right| \le \sigma_Y\right)$$

$$= Prob\left(\frac{2\sqrt{3}}{7}\sigma_Y a \le |P| \le \frac{2\sqrt{3}}{7}\sigma_Y a\right)$$

$$\frac{2\sqrt{3}}{7}\sigma_Y a$$

$$\int f_P(p)dp$$

$$-\frac{2\sqrt{3}}{7}\sigma_Y a$$

(2.9.b) Stresses in all bar are equal, and

$$\sum = \frac{P}{A}; \; R = Prob\left(\left|\sum\right| < \sigma_Y\right) = Prob\left(\frac{|P|}{A} \leq \sigma_Y\right)$$

$$= \int \sigma_Y a \, f_P(p)dp$$

$$-\sigma_Y a$$

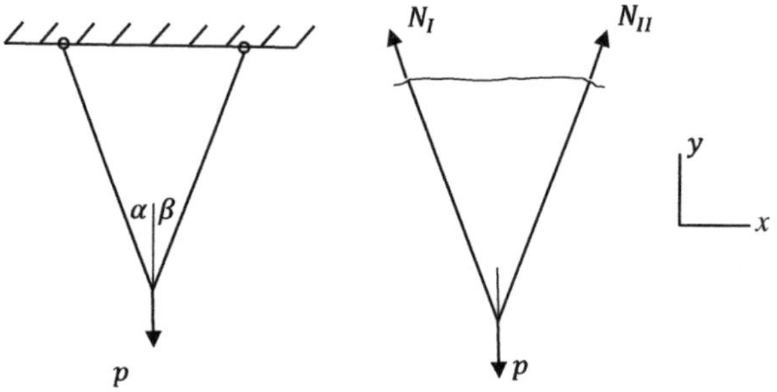

$$\sum^{\rightarrow} X = 0 \; N_2 \cos(90 - \beta) - N_1 \cos(90 - \alpha) = 0 \; N_2 \sin\beta = N_1 \sin\alpha;$$

$$N_2 = N_1 \frac{\sin\alpha}{\sin\beta} \sum Y = 0 \; N_1 \cos\alpha + N_2 \cos\beta$$

$$= P \; N_1 \left(\cos\alpha + \frac{\sin\alpha}{\sin\beta}\cos\beta\right) = P$$

$$N_1 = \frac{P \sin\beta}{\sin(\alpha + \beta)}, \; N_2 = \frac{P \sin\alpha}{\sin(\alpha + \beta)}$$

Now if $\alpha > \beta$, then

$$N_{max} = N_2 = \frac{P \sin\alpha}{\sin(\alpha + \beta)}$$

and

$$R = \text{Prob}\left(\frac{|N_{max}|}{a} \leq \sigma_Y\right) = \int \frac{\sin(\alpha+\beta)}{\sin\alpha}\sigma_Y a - \frac{\sin(\alpha+\beta)}{\sin\alpha}\sigma_Y a\ f_p(p)dp$$

If however, $\alpha < \beta$ then

$$N_{\max} = N_1 = \frac{P\sin\beta}{\sin(\alpha+\beta)}$$

and

$$R = \int \frac{\sin(\alpha+\beta)}{\sin\alpha}\sigma_Y a - \frac{\sin(\alpha+\beta)}{\sin\alpha}\sigma_Y a\ \ f_p(p)dp$$

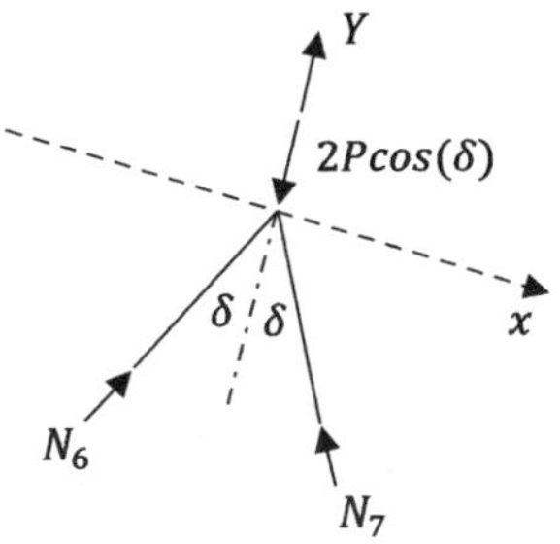

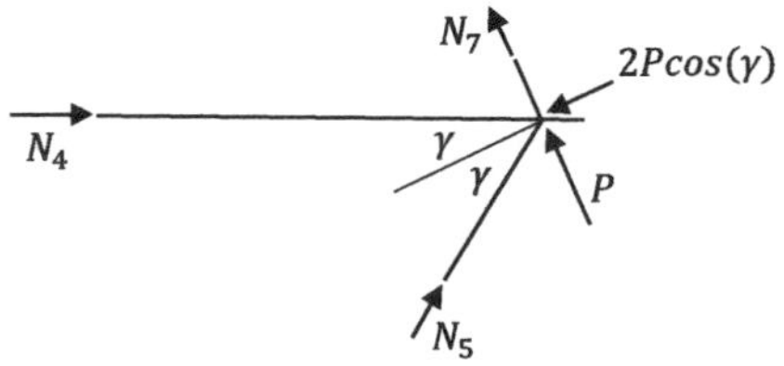

$$\sum^{\rightarrow} X = 0 - N_7\sin\delta + N_6\sin\delta = 0\ N_7 = N_6$$

$$\uparrow \sum = 0 - 2P\cos\delta + N_7\cos\delta + N_6\cos\delta = 0$$

$$-2P\cos\delta + 2N_6\cos\delta = 0\ N_6 = N_7 = P$$

Likewise here

$$N_5 = N_4 = -P$$

In complete analogy, we demonstrate that

$$N_j = P,\ j = 1, 2, \ldots, 7$$

Therefore, the system in figure (2.9.e) is equivalent to that in figure (2.9.b), with the force P directed upward

$$P = \left(\left| \frac{-P}{a} \right| \leq \sigma_Y \right) = \int_{-\sigma_Y a}^{\sigma_Y a} f_P(p)dp$$

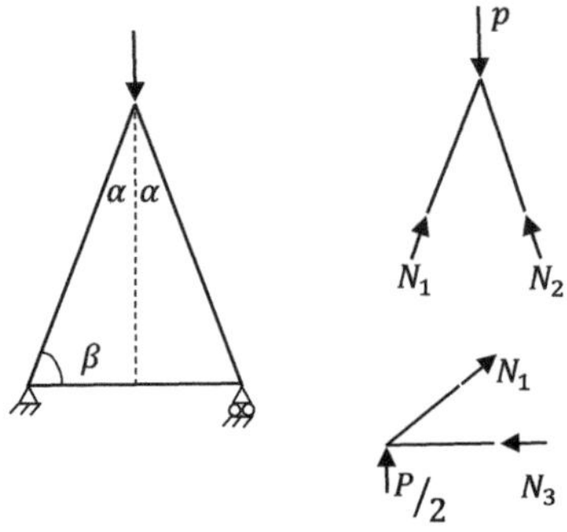

$$\sum X = 0 \quad N_1 = N_2 \uparrow \sum Y = 0; 2N_1 \cos\alpha = P$$

$$N_1 = N_2 = \frac{P}{2\cos\alpha} \sum^{\rightarrow} X = 0 : N_1 \cdot \cos\beta - N_3 = 0$$

$$N_3 = N_1 \cos\beta = \frac{P \sin\alpha}{2\cos\alpha}$$

Let $\alpha = 30°$ [lecturer should specify this angle]. Then:

$$N_1 = \frac{P}{2\cos 30} = \frac{P}{\sqrt{3}} \quad N_3 = \frac{P}{2\sqrt{3}} \quad N_{\max} = N_1 = \frac{P}{\sqrt{3}} \quad \sum = \frac{P}{\sqrt{3}a}$$

$$R = \text{Prob}\left(\frac{|P|}{\sqrt{3}a} \leq \sigma_Y \right) = \int_{-\sqrt{3}\sigma_Y a}^{\sqrt{3}\sigma_Y a} f_P(p)dp$$

PROBLEM 5.3

The truss shown in the figure carries the random normally distributed load P with $E(p) = 10$ kips, $\sigma_P = 5$ kips, $a = 10'$, $\sigma_{\text{allow}} = 24,000$ psi, area $= 1\ \text{in}^2$. Check whether the reliability exceeds the desired $r = 0.999$.

SOLUTION 5.3

The following sections are performed to determine the axial forces in the bars:

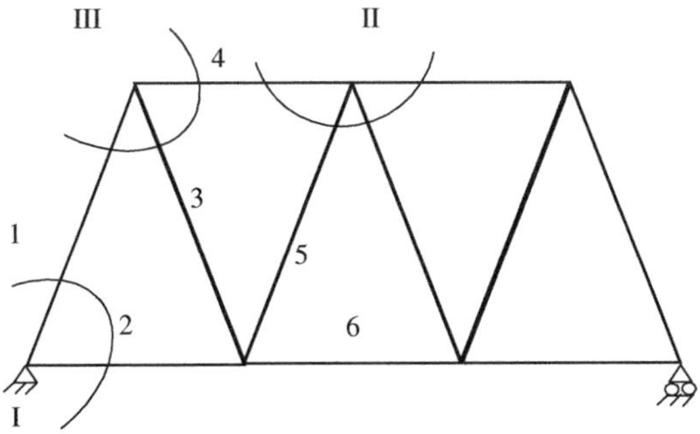

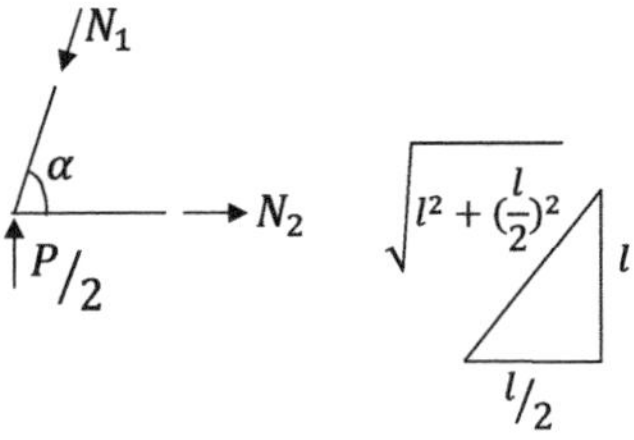

By symmetry, the vertical reactions at the support equals $P/2$; the horizontal reactions vanish.

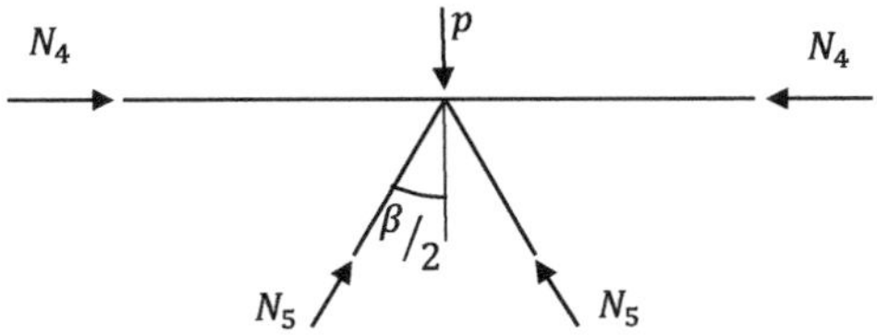

$$\uparrow \sum Y = 0; \; \frac{P}{2} - N_1 \cdot \sin\alpha = 0$$

$$\uparrow \sum Y = 0; \quad \frac{P}{2} - N_1 \cdot \sin\alpha = 0$$

$$N_1 = \frac{P}{2\sin\alpha}; \quad \sin\alpha = \frac{l}{2\sqrt{1+\frac{1}{4}}} = \frac{2}{\sqrt{5}}; \; N = \frac{\sqrt{5}}{4}P$$

$$\cos\alpha = \frac{1}{\sqrt{5}} \quad \sum\nolimits^{\rightarrow} X = 0 \quad N_2 - N_1\cos\alpha = 0, \; N_2 = N_1$$

$$\cos\alpha = \frac{\sqrt{5}}{4}P \cdot \frac{1}{\sqrt{5}} = \frac{P}{4}$$

$$\uparrow \sum Y = 0 - P + 2N_5 \cos\frac{\beta}{2} = 0$$

$$N_5 = \frac{P}{2\cos\beta/2}$$

$$\beta = 180 - 2\alpha; \quad \frac{\beta}{2} = 90 - \alpha; \cos\frac{\beta}{2} = \sin\alpha$$

Therefore:

$$N_5 = \frac{P}{2\cdot(2\sqrt{5})} = \frac{\sqrt{5}}{4}P;$$

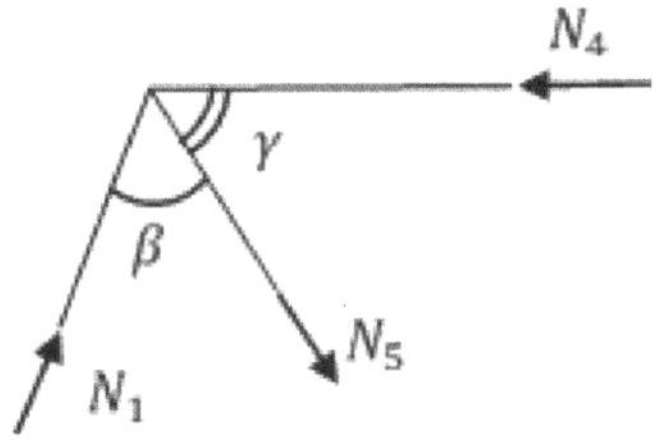

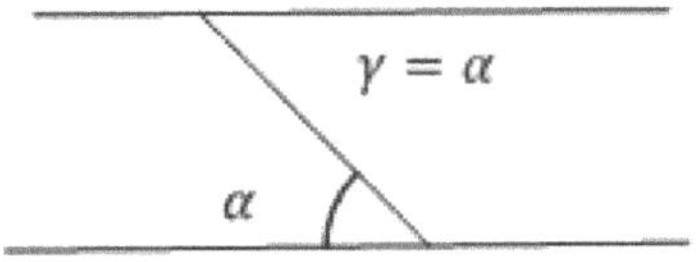

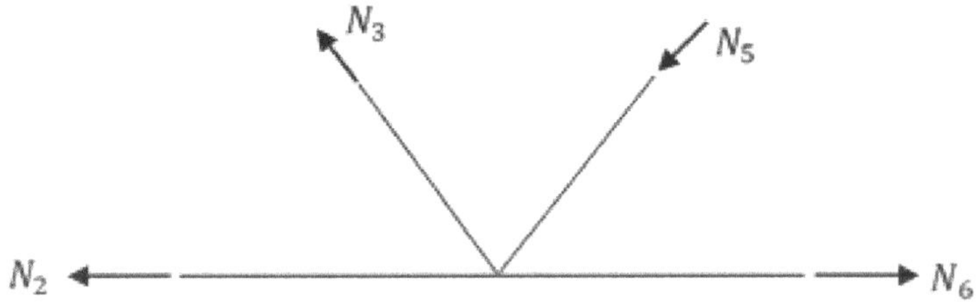

$$\uparrow \sum Y = 0 \quad N_1 \cos(\beta/2) - N_3 Y \cos(\beta/2) = 0; \quad \rightarrow N_3 = N_1 = \frac{\sqrt{5}}{4}P$$

$$\sum\nolimits^{\rightarrow} X = 0; \; -N_4 + N_1 \cos\gamma + N_3 \cos\gamma = 0 \quad N_4 = 2N_1 \cos\gamma$$

$$\gamma = 90^\circ - \frac{\beta}{2} = 90^\circ - \left(90^\circ - \alpha\right) = \alpha$$

$$N_4 = 2 \cdot \frac{\sqrt{5}}{4} P \left(\frac{2}{\sqrt{5}} \right) = P \quad \sum X = 0; \; N_6 - N_2 - (N_3 + N_5) \cos\alpha = 0$$

$$N_6 = N_2 + (N_3 + N_5) \cos\alpha = \frac{P}{4} + \left(\frac{\sqrt{5}}{4} P + \frac{\sqrt{5}}{4} P \right) \cdot \frac{1}{\sqrt{5}}$$

$$= \frac{P}{4} + \left(\frac{P}{4} + \frac{P}{4} \right) = \frac{3}{4} P$$

The results can be put in the table

Bar Number	N_1	sign
1	$-\frac{\sqrt{5}}{4} P$	compression
2	$\frac{1}{4} P$	tension
3	$\frac{\sqrt{5}}{4} P$	tension
4	$-P$	compression
5	$\frac{\sqrt{5}}{4} P$	compression
6	$\frac{3}{4}$	tension

Now, the maximal absolute value is obtained for bars 1, 3, 5, and 6. Therefore,

$$R = \text{Prob}\left(\frac{\sqrt{5}}{4} \frac{|P|}{a} \le \sigma_Y \right) = \int_{-\frac{4}{\sqrt{5}}}^{\frac{4}{\sqrt{5}}} f_P(p) dp,$$

$$= F_P\left(\frac{4}{\sqrt{5}} \sigma_Y a \right) - F_P\left(-\frac{4}{\sqrt{5}} \sigma_Y a \right)$$

Now, since P is normally distributed, Equation 4.13 yields

$$R = \left\{ \frac{1}{2} + \text{erf}\left[\frac{\frac{4}{\sqrt{5}} \sigma_Y a - E(p)}{\sigma_P} \right] \right\} - \left\{ \frac{1}{2} + \text{erf}\left[\frac{-\frac{4}{\sqrt{5}} \sigma_Y a E(p)}{\sigma_P} \right] \right\}$$

$$= \operatorname{erf}\left[\frac{\frac{4}{\sqrt{5}}\sigma_Y a - E(p)}{\sigma_P}\right] + \operatorname{erf}\left[\frac{\frac{4}{\sqrt{5}}\sigma a + E(p)}{\sigma_P}\right]$$

$$= \operatorname{erf}\left(\frac{4}{2.236}\cdot\frac{24\cdot 1}{5} - \frac{10}{5}\right) + \operatorname{erf}\left(\frac{4}{2.236}\cdot\frac{24\cdot 1}{5} + \frac{10}{5}\right)$$

$$= \operatorname{erf}(6.589) + \operatorname{erf}(10.589) \approx 1$$

since $\sigma_Y = 24$ kpsi.

PROBLEM 5.4

A cantilever of rectangular cross section is loaded as shown in the figure. G is a random variable with given $F_G(g)$; σ_Y, the yield stress in the tensile test of the cantilever material, is given. Use the maximum shear-stress criterion to find the reliability of the cantilever. On the concrete numerical example, discuss the change of the reliability estimate under the von Mises criterion.

SOLUTION 5.4

The distributed force acting at the edge of cantilever is statistically equivalent to the same distributed force acting along its middle plus a distributed torsion moment as shown in the figure

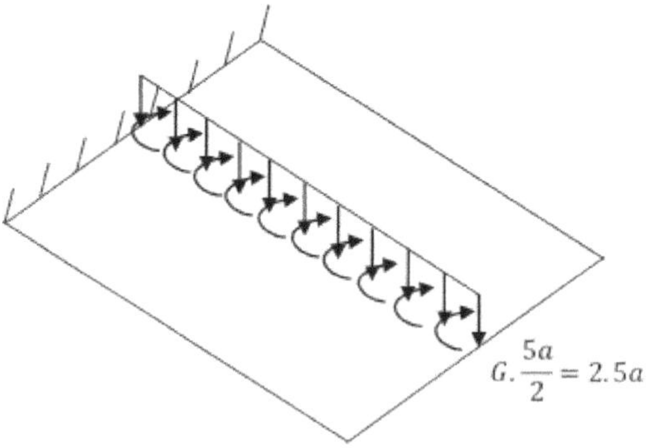

The maximum bending moment equals

$$M_{z,max} = \frac{G\cdot(30a)}{2} = 450Ga^2$$

The maximum torsion equals

$$T_{max} = 2.5\,Ga\cdot 30a = 75\,Ga^2$$

The maximum normal and shear stresses occur at the clamped edge, at the midpoints of the horizontal sides of the rectangle as shown in the Figure (points A and B).

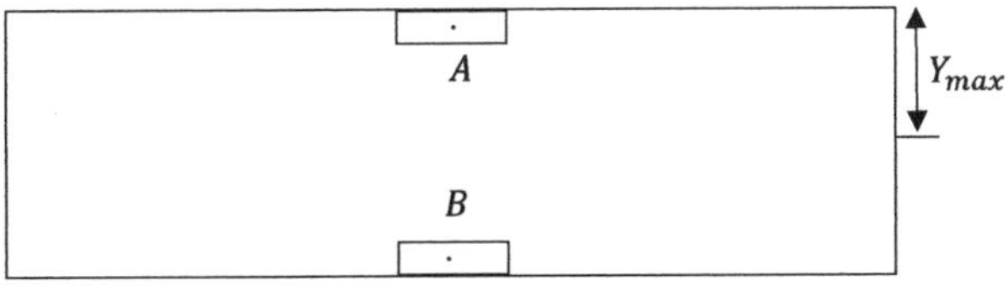

$$\sigma = \frac{M_{z,max}}{I_z} y_{max} = \frac{450\,Ga^2}{\frac{5a\cdot a^3}{12}} \cdot \frac{a}{2} = 540\frac{G}{a}$$

The maximum shear stress due to torsion (see Popov, "Introduction to Mechanics of Solides, McGraw Hill, 1975, p. 167; or Crandall et al, "An Introduction to the Mechanics of Solids", McGraw Hill, 1969, p. 395) is

$$\tau_T = \frac{T}{\alpha \cdot 5a^3}$$

since the side ratio is $5a/a = 5$, $\alpha = 1/3$, yielding,

$$\tau_T = \frac{75\,Ga^2}{\frac{1}{3}\cdot 5a^3} = 45\frac{G}{a} \quad \tau = \frac{T}{wt^2}\left(3 + 1.8\frac{t}{w}\right):$$

According to the maximum shear stress criterion, we need an expression for the maximum shear stress on the element:

$$\tau_{max} = \sqrt{\left(\frac{\sigma}{2}\right)^2 + \tau_T^2}$$

This stress should be compared with its counterpart in the element under uniaxial tension, under the stress equal σ_Y (with σ_Y denoting the yield stress)

$$\tau = \frac{\sigma_Y}{2}$$

so that the strength criterion reads

$$\tau_{max} = \sqrt{\left(\frac{\sigma}{2}\right)^2 + \tau_T^2} \leq \frac{\sigma_Y}{2}$$

or

$$\sigma_{eq} = \sqrt{\sigma^2 + 4\tau_T^2} \leq \sigma_Y$$

σ_{eq} denoting an equivalent normal stress.

Now

$$\sigma_{eq} = \sqrt{\left(540\frac{G}{a}\right)^2 + 4\left(\frac{45G}{a}\right)^2}$$

$$= \sqrt{2 \cdot 270\frac{G}{a})^2 + 4\left(\frac{45G}{a}\right)^2} = 2 \cdot \sqrt{(270)^2 + (45)^2}\left(\frac{G}{a}\right)$$

$$= 90\sqrt{6^2+1}\frac{G}{a} = 90\sqrt{37}\frac{G}{a} = 547.45\frac{G}{a}$$

and the reliability is

$$R = Prob\left(\sigma_{eq} < \sigma_Y\right) = P\left(547.45\frac{G}{a} < \sigma_Y\right) = F_G\left(\frac{\sigma_Y a}{547.45}\right)$$

we next discuss the change of reliability estimate under the Von Mises criterion (maximum distortion energy criterion)

$$\sigma_{eq} = \frac{1}{\sqrt{2}}\sqrt{(\sigma_1-\sigma_2)^2 + (\sigma_2-\sigma_3)^2 + (\sigma_3-\sigma_1)^2}$$

Where we have σ_1, σ_2 and σ_3 are the principal stresses. Since $\sigma_3 = 0$, we have

$$\sigma_{eq} = \sqrt{\sigma_1^2 + \sigma_2^2 - \sigma_1\sigma_2}$$

Now,

$$\sigma_{1,2} = \frac{\sigma}{2} \pm \sqrt{\left(\frac{\sigma}{2}\right)^2 + \tau_T^2}$$

or, denoting

$$\frac{\sigma}{2} = a, \quad \sqrt{\left(\frac{\sigma}{2}\right)^2 + \tau_T^2} = b$$

$$\sigma_{1,2} = a \pm b$$

Then

$$\sigma_{eq} = \sqrt{(a+b)^2 + (a-b)^2 - (a+b)(a-b)}$$

$$= \sqrt{a^2+3b^2}$$

or

$$\sigma_{eq} = \sqrt{\left(\frac{\sigma}{2}\right)^2 + 3\left[\left(\frac{\sigma}{2}\right)^2 + \tau_T^2\right]}$$

$$= \sqrt{\sigma^2 + 3\tau_T^2}$$

substituting

$$\sigma = 450\frac{G}{a} \quad \text{and} \quad \tau_T = 45\frac{G}{a}$$

we find

$$\sigma_{eq} = \sqrt{(540)^2 + 3(45)^2}\,\frac{G}{a} = 45\sqrt{12^2 + 3}\frac{G}{a}$$

$$= 545.6\frac{G}{a}$$

Therefore the reliability is

$$R = Prob\left(\sigma_{eq} \le \sigma_Y\right) = P\left(545.6 \cdot \frac{G}{a} \le \sigma_Y\right) = F_G\left(\frac{\sigma_Y a}{545.6}\right)$$

The argument of the probability distribution function, according to Von Mises theory exceeds that according to the maximum shear stress theory:

$$\frac{\sigma_Y a}{545.6} > \frac{\sigma_Y a}{547.45}$$

since the probability distribution function is a nondecreasing one. The reliability estimate according to the Von Mises criterion is not less than that obtained in the maximum shear stress theory.

Numerical example: Let G have an exponential distribution

$$F_G\left(g\right) = \left(1 - e^{-\alpha g}\right)\, U(g)$$

Then according to the maximum shear stress theory:

$$R_1 = 1 - \exp\left(\frac{-\alpha}{457.45}\sigma_Y a\right)$$

and according to the Von Mises criterion

$$R_2 = 1 - \exp\left(\frac{\alpha}{-545.6}\sigma_Y a\right)$$

Let

$$\sigma_Y = 24\,kN/cm^2 \quad a = 10\,cm \quad \alpha = 0.005\,cm/N = 5\,cm/kN$$

Then

$$R_1 = 1 - \exp\left(-\frac{5}{547.6}24 \cdot 10\right) = 1 - e^{-2.192} = 1 - 0.1117 = 0.8883$$

$$R_2 = 1 - \exp\left(-\frac{5}{545.6}24 \cdot 10\right) = 1 - e^{-2.199} = 1 - 0.1109 = 0.8891 > R_1$$

In addition, let us check whether the strength requirement in the mean is met:

$$E(G) = \frac{1}{\alpha} = 0.2\,kN/cm^2$$

Then according to the maximum shear stress theory,

$$E_{(1)}\left(\sigma_{eq}\right) = 547.45\frac{E(G)}{a} = 547.45\frac{0.2}{10} = 10.95\,kN/cm^2$$

and according to the Von Mises criterion

$$E_{(2)}\left(\sigma_{eq}\right) = 545.6\frac{E(G)}{a} = 545.6\frac{0.2}{10} = 10.91\,kN/cm^2$$

Since

$$E_{(j)}\left(\sigma_{eq}\right) < \sigma_Y = 24\,kN/cm^2, (j = 1, 2)$$

we conclude that the strength requirement in the mean is met according to both theories.

PROBLEM 5.5

A beam is loaded as shown in the figure. G is a random variable with given probability distribution function $F_G(g)$ and α is a given number. Verify that the extremal bending moment occurs at the section $x = (5\alpha - 1)/4\alpha$ and equals

$$M_z = Ga^2\left[\left(\frac{3+\alpha}{4}\right)\left(\frac{5\alpha-1}{4\alpha}\right) - \frac{1}{2}\left(\frac{5\alpha-1}{4\alpha}\right)^2 - \left(\frac{\alpha-1}{2}\right)\left(\frac{\alpha-1}{4\alpha}\right)^2\right]$$

and derive the reliability.

SOLUTION 5.5

Note to lecturer: x should be $x = \frac{5\alpha-1}{4\alpha}a$

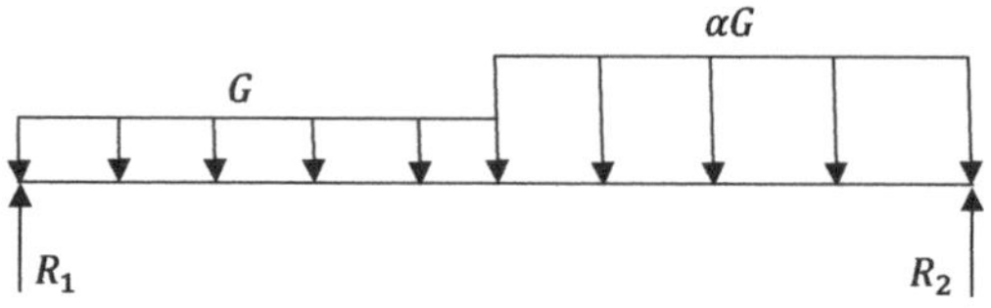

$$\sum M_2 = 0 R_1 \cdot 2a\,(Ga) \cdot 1.5a - \alpha Ga \cdot 0.5a = 0$$

$$R_1 = \frac{(1.5 + 0.5\alpha)\,Ga}{2} = \frac{(3+\alpha)\,Ga}{4}$$

$$\sum M_1 = 0\; R_2 \cdot 2a - Ga \cdot 0.5a - \alpha G \cdot a \cdot 1.5a = 0$$

$$R_2 = \frac{(0.5 + 1.5\alpha)\,Ga}{2} = \frac{(1+3\alpha)\,Ga}{4}$$

Now

$$M_z(x) = \begin{cases} R_1 x - \dfrac{Gx^2}{2}, & 0 \le x \le a \\ R_2 y - \dfrac{\alpha G \cdot y^2}{2}, & 0 \le y \le a \end{cases}$$

$M_{z\,max}$ in the first region

$$M_z^{'}(x) = R_1 - Gx^* = 0,\; x^* = \frac{R_1}{G} = \frac{(3+\alpha)\,a}{4}$$

and has the sense if $\alpha \le 1$ (otherwise the maximum would appear in the second region $x > a$, whereas we use the expression of $M_z(x)$ in the first region).

For $\alpha \le 1$

$$M_z\left(x^*\right) = M_z|x^* = \frac{(3+\alpha)\,a}{4} = \frac{(3+\alpha)\,Ga}{4} \cdot \frac{(3+\alpha)\,a}{4} - \frac{G}{2} \cdot \frac{(3+\alpha)^2 a^2}{16}$$

$$= \frac{(3+\alpha)^2 Ga^2}{16} - \frac{(3+\alpha)^2\,Ga^2}{32} = \frac{(3+\alpha)^2 Ga^2}{32}$$

In the second region:

$$M_z^{'}(y) = R_2 - \alpha Gy^* = 0;\; y^* = \frac{R_2}{\alpha G} = \frac{(1+3\alpha)\,Ga}{4\alpha G} = \frac{1+3\alpha}{4\alpha} a$$

valid for $\alpha \ge 1$.

This value corresponds to that mentioned in the monograph. Indeed, the x coordinate associated with y^*

$$x = 2a - \frac{1+3\alpha}{4\alpha} a = \frac{5\alpha - 1}{4\alpha} a$$

$$M_z\left(y^*\right) = R_2 \cdot \frac{1+3\alpha}{4\alpha} a - \frac{\alpha G}{2} \frac{(1+3\alpha)^2}{16\alpha^2} a^2$$

$$= \frac{(1+3\alpha)^2}{16\alpha} Ga^2 - \frac{(1+3\alpha)^2}{32\alpha} Ga^2 = \frac{(1+3\alpha)^2}{32\alpha} Ga^2$$

Now:

$$\max M_z = \begin{cases} \dfrac{(3+\alpha)^2 G a^2}{32}, & \text{for } \alpha \leq 1 \\ \dfrac{(1+3\alpha)^2 G a^2}{32\alpha}, & \text{for } \alpha \geq 1 \end{cases}$$

The reliability is given by

$$R = Prob\left(\left|\sum\right| < \sigma_y\right) = Prob(z|G| \leq \sigma_Y)$$

where

$$z = \begin{cases} \dfrac{(3+\alpha)^2 a^2}{32}, & \text{for } \alpha \leq 1 \\ \dfrac{(1+3\alpha)^2 a^2}{32\alpha}, & \text{for } \alpha \geq 1 \end{cases}$$

Therefore

$$R = Prob\left(|G| \leq \frac{\sigma_Y}{z}\right) = Prob\left(-\frac{\sigma_Y}{z} \leq G \leq \frac{\sigma_Y}{z}\right)$$

$$= F_G\left(\frac{\sigma_Y}{z}\right) - F_G\left(\frac{-\sigma_Y}{z}\right)$$

and finally

$$R = \begin{cases} F_G\left[\dfrac{32\sigma_Y}{(3+\alpha)^2 a^2}\right] - F_G\left[-\dfrac{32\sigma_Y}{(3+\alpha)^2 a^2}\right], & \text{for } \alpha \leq 1 \\ F_G\left[\dfrac{32\alpha\sigma_Y}{(1+3\alpha)^2 a^2}\right] - F_G\left[-\dfrac{32\alpha\sigma_Y}{(1+3\alpha)^2 a^2}\right], & \text{for } \alpha \geq 1 \end{cases}$$

in the particular case $\alpha = 1$, i.e. beam of span $2a$, under uniform distributed load,

$$R = F_G\left[\frac{2\sigma_Y}{a^2}\right] - F_G\left[-\frac{2\sigma_Y}{a^2}\right]$$

PROBLEM 5.6

Determine the reliability of the cantilever (see figure) under a given force q applied at the random distance X from the clamped edge. $F_X(x)$ is given. (For a generalization of this problem with both X and q random variables, or with random variables, or with random concentrated loads and moments applied at random positions on the beams with different boundary conditions, see the paper by Shukla and Stark.)

SOLUTION 5.6

The external moment occurs at the clamped edge

$$M_z = qX$$

The stress, which is likewise a random variable

$$\sum = \frac{qX}{S}$$

where S is the section modulus.

$$R = \text{Prob}\left(\left|\sum\right| < \sigma_Y\right) = \text{Prob}\left(\frac{|qX|}{S} \leq \sigma_Y\right)$$

$$= Prob\left(\frac{|q|X}{S} \leq \sigma_Y\right) = \text{Prob}\left(\frac{|qX|}{S} \leq \sigma_Y\right)$$

$$= Prob\left(\frac{|q|X}{S} \leq \sigma_Y\right) \text{ since } X \geq 0$$

and

$$R = \text{Prob}\left(X < \frac{\sigma_Y S}{|q|}\right) = F_X\left(\frac{\sigma_Y S}{|q|}\right)$$

PROBLEM 5.7

A thick-walled cylinder (see figure) is under external pressure P with a discrete uniform distribution

$$F_P(p) = \frac{1}{10}\sum_{i=1}^{10} U(x - p_0 i)$$

that is, P takes on values $p_0, 2p_0, \ldots, 10p_0$ with constant probability 1/10. For the transverse stresses, the following expressions are valid (see, e.g., Timoshenko and Goodier):

$$\sum_r = P\frac{(r_o/r_i)^2 - (r_o/r)^2}{(r_o/r_i)^2} \qquad \sum_\theta = -P\frac{(r_o/r_i)^2 + (r_o/r)^2}{(r_o/r_i)^2 - 1}$$

where r_o and r_i and the outer and inner radii, respectively. Using the von Mises criterion, find r_o/r_i such that the desired reliability is not less than 0.99.

Note to lecturer: A minus sign is needed in the equation for $\sum r$. Indeed, for $r = r_0$ we should have $\sum = -P$.

SOLUTION 5.7

$$\sigma_{eq} = \sqrt{-\frac{1}{2}\left[(\sigma_1 - \sigma_2)^2 + (\sigma_2 - \sigma_3)^2 + (\sigma_3 - \sigma_1)^2\right]}$$

The equation for $\sum$ and $\sum \theta$ can be put in the common form

$$\sum r, \theta = -\frac{Pr_0^2}{r_0^2 - r_i^2}\left(1 \mp \frac{r_i^2}{r^2}\right)$$

The graph of the change of stresses are shown here

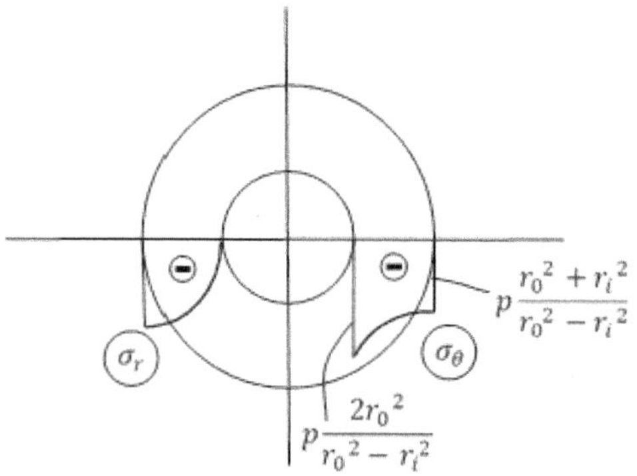

For the inner surface of the cylinder we have:

$$\sigma_1 = 0 \, \sigma_2 = 0 \, \sigma_3 = -P\frac{2r_0^2}{r_0^2 - r_i^2}$$

and

$$\sigma_{eq} = \sqrt{\frac{1}{2}\left[\sigma_2^2 + (\sigma_2 - \sigma_3)^2 + \sigma_3^2\right]} = \sqrt{\frac{1}{2}(\sigma_2^2 + 2\sigma_3^2 - 2\sigma_2\sigma_3)}$$

$$= \sqrt{(\sigma_2 - \sigma_3)^2 + \sigma_2\sigma_3};$$

$$\sigma_2 - \sigma_3 = -P + P\left(r_0^2 + r_i^2\right)\left(r_0^2 - r_i^2\right) = P \cdot 2r_i^2/(r_0^2 - r_i^2);$$

$$\sigma_{eq} = \sqrt{P^2 \cdot \frac{4r_1^4}{\left(r_0^2 - r_i^2\right)^2} + P^2\frac{r_0^2 + r_i^2}{r_0^2 - r_i^2}} = \frac{|P|}{r_0^2 - r_i^2}\sqrt{r_0^4 + 3r_i^4}$$

Comparison of the equivalent stresses at the two surfaces shows that the maximum value occurs at the inner surface.

For $P_0 > 0$,

$$R = \text{Prob}\left(P\frac{2r_0^2}{r_0^2 - r_i^2} \le \sigma_Y\right)$$

$$= F_P\left(\sigma_Y \frac{r_0^2 - r_i^2}{2r_0^2}\right)$$

Now the graph of $F_P(p)$ is

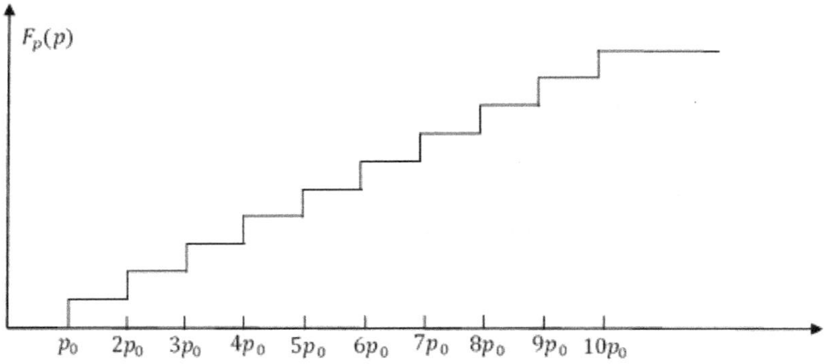

we need, that

$$R \ge 0.99$$

this means, for our discrete random variable

$$F_P\left(\sigma_Y \frac{r_0^2 - r_i^2}{2r_0^2}\right) = 1,$$

or

$$F_P\left[\sigma_Y\left(\frac{1}{2} - \frac{1}{2}\frac{r_i^2}{r_0^2}\right)\right] = 1$$

And

$$\sigma_Y\left(\frac{1}{2} - \frac{1}{2}\frac{r_i^2}{r_0^2}\right) \ge 10P_0$$

$$1 - \frac{r_i^2}{r_0^2} \ge 20\frac{P_0}{\sigma_Y}$$

$$\frac{r_i^2}{r_0^2} \le 1 - 20\frac{P_0}{\sigma_Y}$$

PROBLEM 5.8

A rectangular plate, simply supported all around, is under a load Q uniform over its surface, with chi-square distribution (Sec. 4.7)

$$f_Q(q) = \frac{e^{-q/2} q^{\nu-1}}{2^{\nu}\Gamma(\nu)},\ q > 0,\ \nu = \frac{m}{2}$$

The displacement of the plate under a deterministic uniform load q is (see, e.g., Timoshenko and Woinowski-Krieger):

$$w = \frac{16q}{\pi^2 D} \sum_{m=1}^{\infty} \sum_{n=1}^{\infty} \frac{1}{mn(m^2/a^2 + n^2/b^2)^2} \sin\frac{m\pi x}{a} \sin\frac{n\pi y}{b}$$

where a and b are the sides of the plate; $D = Eh^3/12(1 - \nu^2)$ is the flexural rigidity; E, the modulus of elasticity; ν, Poisson's ratio; and h the thickness of the plate. Find the probability density of the maximum deflection.

1 The note to problem 5.1 applies here as well.
2 π in the formula for w is the sixth power, not the second.

SOLUTION 5.8

The maximum deflection of the plate occurs at some point say (x_0, y_0) depending on the side ratio b/a,

$$y \equiv W_{max} = \frac{16q}{\pi^6 D} \sum_{m=1}^{\infty} \sum_{n=1}^{\infty} \frac{1}{mn(m^2/a^2 + n^2/b^2)^2} \sin\frac{mnx_0}{a} \sin\frac{n\pi y_0}{a} = Z_q$$

where

$$Z = \frac{16}{\pi^6 D} \sum_{m=1}^{\infty} \sum_{n=1}^{\infty} \frac{1}{mn(m^2/a^2 + n^2/b^2)^2} \sin\frac{mnx_0}{a} \sin\frac{n\pi y_0}{a}$$

so that we have a linear transformation

$$f_Y(y) = \frac{1}{|Z|} f_Q\left(\frac{y}{Z}\right)$$

according to Equation (4.46). Hence,

$$f_Y(y) = \frac{1}{|Z|} \frac{e^{-y/2Z}(y/Z)^{\nu-1}}{2^{\nu}\Gamma(\nu)},\ \nu = \frac{m}{2}$$

The series for Z is rapidly convergent. Timoshenko and Woinowski-Krieger, on page 120 (Table 8) give the values for α, for rectangular plates,

$$W_{max} = \frac{\alpha q a^4}{D}$$

for various b/a ratios.

For a square plate we have

$$X_0 = \frac{a}{2}, \quad y_0 = \frac{b}{2}$$

$$Z = \frac{16a^4}{\pi^6 D} \cdot \sum_{m=1}^{\infty}\sum_{n=1}^{\infty} \frac{1}{mn(m^2+n^2)^2} \sin\frac{m\pi}{2} \sin\frac{n\pi}{2} = 0.00406$$

hence,

$$Z = 0.00406\frac{a^4}{D}$$

and

$$f_Y(y) = \frac{D}{0.00406a^4} \frac{e^{-\frac{yD}{0.00812}}(yD/0.00812)^{\nu-1}}{2\nu\Gamma(\nu)}, \quad \nu = \frac{m}{2}$$

PROBLEM 5.9

Derive an equation analogous to (5.63) for an asymmetric structure.

SOLUTION 5.9

In the static case the buckling load of an asymmetric structure is governed by Equation (5.53):

$$\left(1 - \frac{\lambda_s}{\lambda_c}\right)^2 + 4a\bar{\xi}\frac{\lambda_s}{\lambda_c} = 0$$

For the dynamic case, we have to use the analogy derived in Equation (5.60). Applying the formula substitution

$$\lambda_s + \lambda_d, \quad a + \frac{2}{3}a, \quad \bar{\xi} + 2\bar{\xi}$$

we have

$$\left(1 - \frac{\lambda_d}{\lambda_s}\right)^2 + 4 \cdot \frac{2}{3}a \cdot 2\bar{\xi}\frac{\lambda_d}{\lambda_s} = 0$$

or

$$\left(1 - \frac{\lambda_d}{\lambda_s}\right)^2 + \frac{16}{3} a\bar{\xi} \frac{\lambda_d}{\lambda_s} = 0 \tag{2}$$

which coincides with the equation which follows Equation (5.62).

Remark

For completeness, let us derive Equation (1). For an asymmetric structure, via Equation (5.45)

$$\frac{\lambda}{\lambda_c} = \frac{\xi + a\xi^2}{\xi + \bar{\xi}} \tag{3}$$

which is equivalent to

$$a\xi^2 + \xi\left(1 - \frac{\lambda}{\lambda_c}\right) - \frac{\lambda}{\lambda_c}\bar{\xi} = 0 \tag{4}$$

Now for finding λ_s, we differentiate (3)

$$\frac{d}{d\xi}\frac{\lambda}{\lambda_c} \frac{(\xi + \bar{\xi})(1 + 2a\bar{\xi}) - (\xi + a\xi^2)}{(\xi + \bar{\xi})^2} = 0 \tag{5}$$

or

$$a\xi^2 = \frac{\lambda}{\lambda_c}\bar{\xi} - \xi\left(1 - \frac{\lambda}{\lambda_c}\right)$$

which, substituted in Equation (5), yields, with λ/λ_s under Equation (5):

$$\frac{\lambda_s}{\lambda_c}\bar{\xi} - \xi\left(1 - \frac{\lambda_s}{\lambda_c}\right) + 2a\bar{\xi}\xi + \bar{\xi} = 0$$

$$\xi = \bar{\xi}\frac{1 + \lambda_s/\lambda_c}{1 - \lambda_s/\lambda_c - 2a\bar{\xi}}$$

At this value max $\lambda = \lambda_s$ occurs. To find the relationship λ_s/λ_c, versus $\bar{\xi}$ we substitute Equation (6) in Equation (5), namely

$$a\bar{\xi}^2 \frac{(1 + \lambda_s/\lambda_c)^2}{(1 - \lambda_s/\lambda_c - 2a\bar{\xi})^2} + 2a\bar{\xi}^2 \frac{1 + \lambda_s/\lambda_c}{1 - \lambda_s/\lambda_c} - 2a\bar{\xi} + \bar{\xi} = 0$$

and simple algebra yields

$$a\bar{\xi}\frac{(1 + \lambda_s/\lambda_c)^2}{(1 - \lambda_s/\lambda_c - 2a\bar{\xi})^2} + 2a\bar{\xi}\frac{1 + \lambda_s/\lambda_c}{1 - \lambda_s/\lambda_c - 2a\bar{\xi}} + 1 = 0$$

$$a\bar{\xi}(1+\lambda_s/\lambda_c)^2 + 2a\bar{\xi}(1+\lambda_s/\lambda_c)(1-\lambda_s/\lambda_c-2a\bar{\xi})$$
$$+(1-\lambda_s/\lambda_c-2a\bar{\xi})^2 = 0$$
$$a\bar{\xi}(1+\lambda_s/\lambda_c)^2 + 2a\bar{\xi}(1-\lambda_s^2/\lambda_c^2) - 4a^2\bar{\xi}^2(1+\lambda_s/\lambda_c)$$
$$+(1-\lambda_s/\lambda_c)^2 + 4a^2\bar{\xi}^2 - 4a\bar{\xi}(1-\lambda_s/\lambda_c) = 0$$

Further:

$$a\bar{\xi}\left(1+2\frac{\lambda_s}{\lambda_c}+\frac{\lambda_s^2}{\lambda_c^2}+2-2\frac{\lambda_s^2}{\lambda_c^2}-4+4\frac{\lambda_s}{\lambda_c}\right)$$
$$-4a^2\bar{\xi}^2\frac{\lambda_s}{\lambda_c}+\left(1-\frac{\lambda_s}{\lambda_c}\right)^2 = 0$$
$$a\bar{\xi}\left(-1+6\frac{\lambda_s}{\lambda_c}-\frac{\lambda_s^2}{\lambda_c^2}\right)-4a^2\bar{\xi}^2\frac{\lambda_s}{\lambda_c}+\left(1-\frac{\lambda_s}{\lambda_c}\right)^2 = 0$$
$$-\bar{\xi}\left(1-\frac{\lambda_s}{\lambda_c}\right)^2+4a\bar{\xi}\frac{\lambda_s}{\lambda_c}-4a^2\bar{\xi}\frac{\lambda_s}{\lambda_c}+\left(1-\frac{\lambda_s}{\lambda_c}\right)^2 = 0$$

Now, grouping similar terms, we find

$$\left(1-\frac{\lambda_s}{\lambda_c}\right)^2(1-a\bar{\xi})+4a\bar{\xi}\frac{\lambda_s}{\lambda_c}(1-a\bar{\xi}) = 0$$

or

$$(1-a\bar{\xi})\left[\left(1-\frac{\lambda_s}{\lambda_c}\right)^2+4a\bar{\xi}\frac{\lambda_s}{\lambda_c}\right] = 0$$

for $a\bar{\xi} \neq 1$, we obtain finally

$$\left(1-\frac{\lambda_s}{\lambda_c}\right)^2+4a\bar{\xi}\frac{\lambda_s}{\lambda_c} = 0 \tag{7}$$

which is the desired result.

Note that Equation (2) is directly obtainable from Equation (5.58), by first expression $\frac{\lambda}{\lambda_s}$ as

$$\frac{\lambda}{\lambda_c} = \frac{\xi_{max}+\frac{2}{3}a\xi_{max^2}}{\xi_{max}+2\bar{\xi}}$$

and then applying condition (5.59)

$$\frac{d}{d\xi_{max}}\frac{\lambda}{\lambda_c} = 0$$

which would be repeating our derivations in Equations (3) through (7) with the above formal substitution

$$\lambda_s + \lambda_d, \quad a \to \frac{2}{3}a, \quad \bar{\xi} \to 2\bar{\xi}.$$

PROBLEM 5.10

Find the probability density function of the dynamic buckling loads and find the mean dynamic buckling load for the symmetric structure.

SOLUTION 5.10

Consider the case $\bar{X}$ to be $N(0, \sigma^2)$. Then

$$R(\alpha) = P\left(\frac{\Lambda_d}{\Lambda_s} > \alpha\right) = P(-\bar{\xi}'' < \bar{X} < \bar{\xi}'')$$

see Equation 5.67. Here

$$\bar{\xi}'' = \frac{2}{3\sqrt{6}}\frac{(1-\alpha)^{3/2}}{\alpha\sqrt{-b}}$$

and

$$R(\alpha) = 2erf\left[\frac{\bar{\xi}''(\alpha)}{\sigma}\right]$$

The probability of the buckling loads is obtained by differentiation

$$f_{\Lambda_d/\Lambda_s}\left(\frac{\lambda_d}{\lambda_s}\right) = -\left.\frac{dR(\alpha)}{d\alpha}\right|_{\alpha=\lambda_d/\lambda_c}$$

which results in

$$f_{\Lambda_d/\Lambda_s}\left(\frac{\lambda_d}{\lambda_s}\right) = \frac{2}{\sqrt{2\pi}}\exp\left[-\frac{\bar{\xi}''(\lambda_d/\lambda_s)}{\sigma}\right]\frac{1}{\sigma}\cdot\frac{2}{3\sqrt{6}}\frac{(1-\lambda_d/\lambda_s)^{1/2}}{(\lambda_d/\lambda_s)\sqrt{-b}}$$

PROBLEM 5.11

Find the reliability function of the asymmetric structure if the initial imperfection is normally distributed $N(0, \sigma^2)$, in the static setting.

SOLUTION 5.11

An asymmetric structure is statically imperfection – sensitive if $a\xi < \bar{0}$, and the buckling load is found from Equation (5.53)

$$\left(1 - \frac{\lambda_s}{\lambda_c}\right)^2 + 4a\bar{\xi}\frac{\lambda_s}{\lambda_c} = 0 \tag{1}$$

the graph of λ_s/λ_c versus $\bar{\xi}$ is shown below.

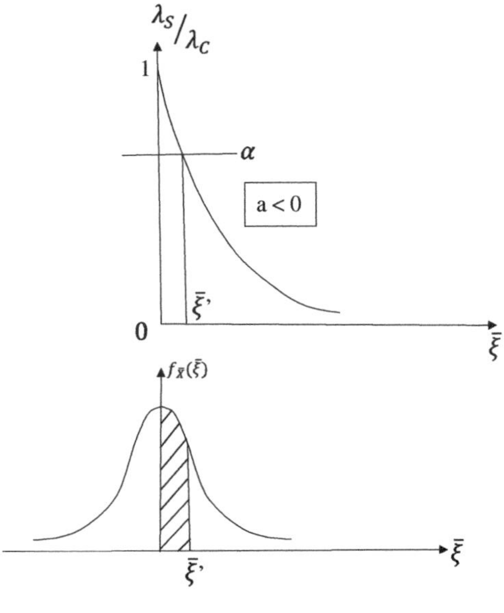

We are interested in the reliability

$$R(\alpha) = P\left(\frac{\Lambda_s}{\lambda_c} > \alpha\right)$$

which implies that

$$R(\alpha) = P(0 \leq \bar{X} \leq \bar{\xi}')$$

The value of $\bar{\xi}'$ is obtained from Equation (1) substituting

$$\lambda_s/\lambda_c = \alpha, \bar{\xi} = \bar{\xi}' :$$

$$(1-\alpha)^2 + 4a\bar{\xi}'\,\alpha = 0$$

or

$$\bar{\xi}' = \frac{(1-\alpha)^2}{-4a\alpha}$$

Since $\bar{\xi} > 0$, $a < 0$ for imperfection — sensitivity, the right-hand side of the latter expression is positive. Thus

$$R(\alpha) = \text{Prob}\left[0 \le \bar{X} \le \frac{(1-\alpha)^2}{-4a\alpha}\right]$$

$$= \text{erf}\left[\frac{(1-\alpha)^2/(-4a\alpha) - E(\bar{X})}{\sigma_{\bar{X}}}\right]$$

$$= \text{erf}\left[\frac{(1-\alpha)^2}{-4a\alpha\sigma_{\bar{X}}}\right]$$

when $\alpha \to 1$, $\text{erf}(0) \to 1/2$ and $R \to 1/2$. This is also understandable from the graphs, since

$$R(1) = \text{prob}(\Lambda_s > 1) = P\left(0 < \bar{X} < \infty\right) = 1/2$$

The reliability curve is shown below.

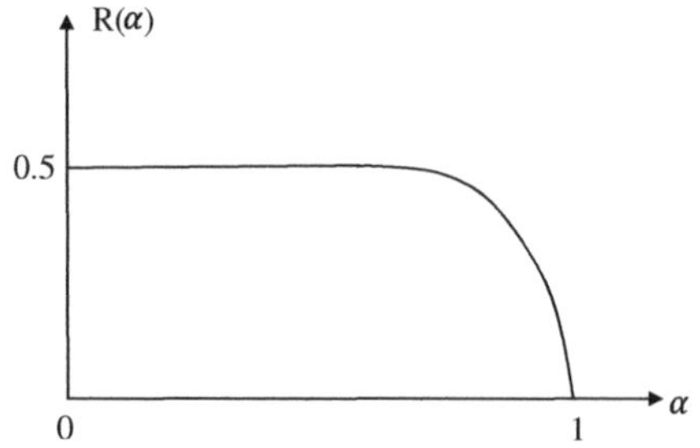

PROBLEM 5.12

Repeat 5.11 for the dynamic buckling problem.

SOLUTION 5.12

An asymmetric structure is dynamically imperfection sensitive of $a\bar{\xi} < 0$, and the buckling load satisfies the equations which follows Equation (5.62):

$$\left(1 - \frac{\lambda_d}{\lambda_c}\right)^2 + \frac{16}{3} a\xi - \frac{\lambda_d}{\lambda_c} = 0$$

The graphs are similar to those for Problem 5.11

$$R(\alpha) = P\left(\frac{\Lambda_d}{\lambda_c} \geq \alpha\right) = \text{Prob}(0 \leq \bar{X} \leq \bar{\xi}')$$

$$= \text{Prob}\left(0 \leq \bar{X} \leq -\frac{(1-\alpha)^2}{\frac{16}{3}a\alpha}\right) = F_{\bar{X}}\left[-\frac{(1-\alpha)^2}{\frac{16}{3}a\alpha}\right]$$

for $\bar{X}$ being $N(0, \sigma_{\bar{X}}^2)$ we get

$$R(\alpha) = erf\left[-\frac{(1-\sigma)^2}{\frac{16}{3}a\sigma_{\bar{X}}}\right]$$

Comparing the reliability expression for the static and dynamic cases, we note that for fixed values of α, a and $\sigma_{\bar{X}}$, the argument of the error function in the dynamic case is lower than its static counterpart. Since the error function is monotonically increasing the reliability is lower in the dynamic case.

PROBLEM 5.13

Assume that the initial imperfection has an exponential distribution, with parameter $E(\bar{X})$ given. Find the reliability of the asymmetric structure.

SOLUTION 5.13

Consider first the static case. Then, as in problem 5.11, we have:

$$\left(1 - \frac{\lambda_s}{\lambda_c}\right)^2 + 4a\bar{\xi}\frac{\lambda_s}{\lambda_c} = 0$$

now, instead of the graphs given there, we have

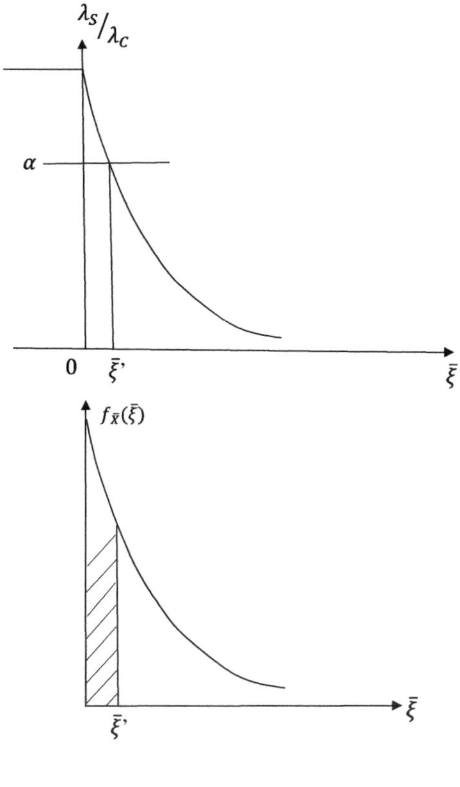

$$R(\alpha) = \text{Prob}(0 \leq \bar{X} \leq \frac{(1-\alpha)^2}{-4a\alpha}$$

$$= \int_0^{\bar{\xi}'} f_X(x)dx$$

for exponential $\bar{X}$, we have

$$f_{\bar{X}}\left(\bar{\xi}\right) = be^{-b\bar{\xi}}$$

according to Equation (4.8); here $b = \frac{1}{E(\bar{X})}$

Hence

$$R\left(\alpha\right) = \int_0^{\bar{\xi}} be^{-bx}dx = 1 - e^{-b\bar{\xi}'}$$

$$= 1 - e^{-\bar{\xi}'/E(\bar{X})}$$

$$= 1 - \exp\left[\frac{(1-\alpha)^2}{4a\alpha E\left(\bar{X}\right)}\right]$$

Since $a < 0$, the argument at the exponent is negative. For $\alpha = 1$

$$R(1) = 0$$

for $\alpha \to 0$, $R(0) \to 1$ unlike the normally-distributed imperfection in Problem 5.11.

For the dynamic case:

$$R(\alpha) = 1 - \exp\left[\frac{(1-\alpha)^2}{\frac{16}{3}a\alpha E\left(\bar{X}\right)}\right]$$

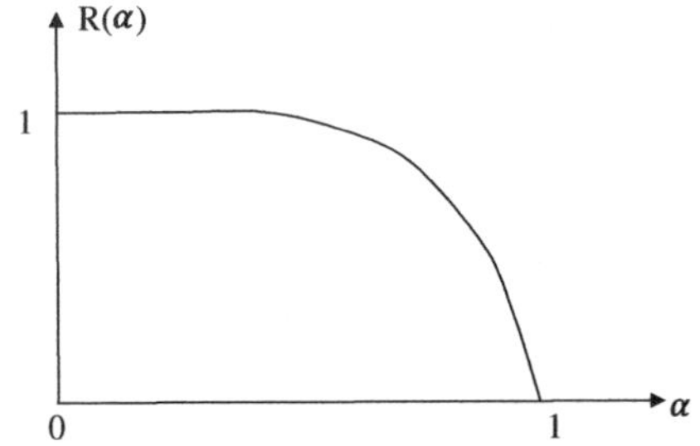

PROBLEM 5.14

Plot the nondimensional static buckling λ_s/λ_c versus initial imperfection $\bar{\xi}$ curve for the nonsymmetric structure, according to Eq. (5.50). Find the reliability at the load level λ.

SOLUTION 5.14

The relationship in Equation (5.50) is depicted below

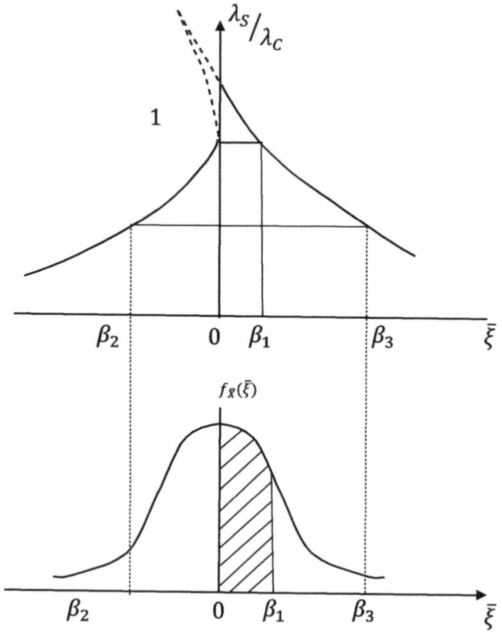

The dashed lines represent the meaningless branches of this equation. Note that the buckling load can exceed that of perfect structure. The probability of this event is (see dashed area)

$$R\,(\alpha = 1) = \mathrm{erf}\left(\frac{\beta_1 - <\bar{\xi}>}{\sigma_{\bar{x}}}\right) + erf\left(\frac{<\bar{\xi}>}{\sigma_{\bar{x}}}\right)$$

Reliability for $\alpha < 1$ is

$$R\,(\alpha) = \mathrm{erf}\left(\frac{\beta_3 - <\bar{\xi}>}{\sigma_{\bar{x}}}\right) - erf\left(\frac{\beta_2 - <\bar{\xi}>}{\sigma_{\bar{x}}}\right)$$

Note that under new circumstances

$$\beta_2 \neq -\beta_3$$

For the reliability of nonsymmetric structures, see the paper entitled, "Reliability Approach to the Random Imperfection Sensitivity of Columns", (by I. Elishakoff), Acta Mechanica, Vol. 55, No. 1–2, 1985, pp. 151–170. The paper considers the buckling of statistically imperfect finite columns on a mixed quadratic-cubic elastic foundation.

PROBLEM 5.15

A rigid weightless bar with a frictionless pin joint at A, constrained by nonlinear springs with $k > 0$, $\beta > 0$, is under an eccentric load P (see figure). The equilibrium equation is

$$P\,(x + \varepsilon) = 2klx(1 - \beta x^2/l^2)$$

yielding $P_c = 2kl$. Find the expression for the maximum force P_{max} supported by the bar as a function of the eccentricity ε. Assume eccentricity to be a continuous random variable with probability density $f_E(\varepsilon)$. Find the reliability of the structure at load level λ.

SOLUTION 5.15

For a perfect (no eccentricity) and linear structure, equilibrium dictates

$$P \cdot x - 2kl \cdot x = 0$$

or

$$(P - 2kl)x = 0$$

This implies that either $x = 0$ and the structure retains its initial straight position (in which case P can be arbitrary), or $x \neq 0$ and

$$P = 2kl = P_{cl}$$

which is the classical buckling load.

For the structure with eccentricity and a nonlinear spring, the equilibrium equation reads:

$$P(x + \varepsilon) = 2klx(1 - \beta x^2/l^2) \tag{2}$$

or

$$\frac{P}{P_{cl}} = \frac{x(1 - \beta x^2/l^2)}{x + \varepsilon} \tag{3}$$

This is equivalent [as is seen in comparison with Equation (5.45)] to the asymmetric structure of Section 5.5, with

$$\bar{\xi} = \varepsilon,\ b = -\frac{\beta}{l^2}$$

Therefore, the maximum load the structure in Problem 5.15 can attain, is obtainable from Equation (5.51) with Equation (4) in mind:

$$\left(1 - \frac{P_s}{P_{cl}}\right)^{\frac{3}{2}} - \frac{3\sqrt{3}}{2}\,|\varepsilon|\,\frac{\sqrt{\beta}}{l^2}\frac{P_s}{P_{cl}} = 0$$

The reliability at the load level λ is:

$$R(\lambda) = Prob\,(P_s > \lambda) = Prob\left(-\varepsilon' \le \varepsilon \le \varepsilon'\right)$$

$$= Prob\left(-\frac{2}{\sqrt{3}}\frac{(1-\lambda/P_{cl})^{\frac{3}{2}}}{(\lambda/P_{cl})\sqrt{\beta/l}} \le \varepsilon \le \frac{2}{\sqrt{3}}\frac{(1-\lambda/P_{cl})^{\frac{3}{2}}}{(\lambda/P_{cl})\sqrt{\beta/l}}\right)$$

$$= F_{\varepsilon}\left[\frac{2}{3\sqrt{3}}\frac{(1-\lambda/P_{cl})^{\frac{3}{2}}}{(\lambda/P_{cl})\sqrt{\beta/l}}\right] - F_{\varepsilon}\left[-\frac{2}{3\sqrt{3}}\frac{(1-\lambda/P_{cl})^{\frac{3}{2}}}{(\lambda/P_{cl})\sqrt{\beta/l}}\right]$$

PROBLEM 5.16

Generalize the results of Sec. 5.7 for the case where the load function is a rectangular impulse $P(t) = P[U(t) - U(t-\tau)]$ with P and τ given positive quantities. (See figure.)

SOLUTION 5.16

Load function in a form of a rectangular impulse can be represented as application of the force P at $t = 0$ and application of the negative force $-P$ at $t = \tau$:

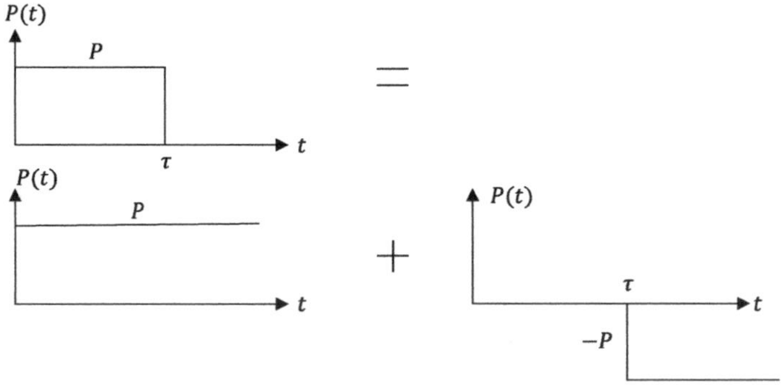

First of these loadings results in displacements given by Equation 5.93 Consider

$$V(\lambda) = \frac{G}{1-\beta}(\cosh r\lambda - \beta)$$

whereas the second loading, for $\lambda > \omega_1\tau \equiv T$ results in the displacement

$$V(\lambda) = \frac{G}{1+\beta}[\cosh r\,(\lambda - T) + \beta]$$

Thus the full displacement will be, due to both loadings

$$V(\lambda) = G\Psi(\lambda)$$

with

$$\Psi(\lambda) = \frac{1}{1-\beta}(\cosh r\lambda - \beta) + \frac{1}{1+\beta}[\cosh r(\lambda - T) + \beta]$$

Further steps are as on pp. 151–152.

PROBLEM 5.17

Generalize the results of Sec. 5.7 for the case where failure is considered in the finite time interval $0 \le t \le t^*$ [see Eq. (5.85)].

SOLUTION 5.17

Consider particular case $\beta = 1$ and G being exponentially distributed, so that

$$V(\lambda) = \frac{1}{2}G(\lambda^2 + 2)$$

and

$$\begin{aligned} R(\lambda, \lambda^*) &= Prob\left[V(\lambda) \le c, 0 \le \lambda \le \lambda^*\right] \\ &= Prob\left[\frac{1}{2}G\left(\lambda^2 + 2\right) \le c,\ 0 \le \lambda \le \lambda^*\right] \\ &= Prob\left[G < \frac{2c}{\lambda^2 + 2},\ 0 \le \lambda \le \lambda^*\right] \end{aligned}$$

where $\lambda^* = \omega_1 t^*$.

Now for $\lambda \le \lambda^*$

$$R(\lambda, \lambda^*) = F_G\left(\frac{2c}{\lambda^2 - 2}\right) = 1 - \exp\left[-\frac{2c}{(\lambda^2 - 2)\,E(G)}\right]$$

This means that reliability can not be smaller than

$$R(\lambda^*, \lambda^*) = 1 - \exp\left[-\frac{2c}{(\lambda^{2*} - 2)\,E(G)}\right]$$

PROBLEM 5.18

Verify that the buckling time Λ in Sec. 5.7 for $P < P_c$ does not represent a random variable. Assign an infinite buckling time to the structure that does not buckle. How does Eq. (5.84) change in these circumstances? Show analytically that for $P < P_c$, $E(\Lambda)$ approaches infinity.

SOLUTION 5.18

For $P < P_c, \beta = 1/\alpha = P_c/P > 1$. This case is considered on p. 153. Now, as is seen, $\{L\}$ is no longer a certain event, maximum displacement might not reach the critical point c.

As we had defined in Section 3.1, the probability $P(\Lambda = +\infty)$ should be zero for Λ to be a random variable. But buckling load being infinity, or structure not failing at all, has a finite probability, for cases depicted on Figure 5.37; these probabilities are $1 - 0.733 = 0.267$ and $1 - 0.555 = 0.454$, for curves 1 and 2, respectively.

We will assign an infinite buckling time to the structure that does not buckle. In large number of structures, there are for data for curve 1 on Figure 5.37 which does not buckle is 26.7%, so that the mean buckling time becomes infinity also.

Let us show this analytically. Indeed

$$E(\Lambda) = \int_0^\infty \lambda \frac{d}{d\lambda} Prob\,(\Lambda \le \lambda)\, d\lambda$$

Integration by parts yields formula (5.119)

$$E(\Lambda) = -\lambda R(\lambda)]_0^\infty + \int_0^\infty R(\lambda)\, d\lambda$$

Now the first term is not zero

$$lim \lambda R(\lambda) \to \infty \quad \text{with } \lambda \to \infty$$

since

$$R(\lambda) \to 0, \quad \lambda \to \infty$$

this yields in infinite mean buckling time, obtained above by intuitive arguments.

PROBLEM 5.19

Modify the results of Sec. 5.7 for the case where the structure possesses viscous damping.

SOLUTION 5.19

With viscous damping, Equation (5.86) is replaced by

$$E_I \frac{\partial^4 y}{\partial x^4} + P\frac{\partial^2 y}{\partial x^2} + c\frac{\partial y}{\partial t} + \rho A\frac{\partial^2 y}{\partial t^2} = -P\frac{\partial^2 y}{\partial x^2}$$

Equation (5.) is replaced by

$$\frac{\partial^4 y}{\partial \xi^4} + \pi^2\alpha\frac{\partial^2 y}{\partial \xi^2} + 2\zeta\pi^4\frac{\partial y}{\partial \lambda} + \pi^4\frac{\partial^2 y}{\partial \lambda^2} = -\alpha\pi^2\frac{\partial^2 y}{\partial \xi^2}$$

where in addition to notations on the top of p. 149, we have denoted

$$c = 2\zeta\omega 1\rho A$$

Now with Equations (5.89) and (5.90) in mind, we get instead of 5.91

$$\frac{d^2 e}{d\lambda^2} + 2\zeta\frac{de}{d\lambda} + (1-\alpha)\, e = \alpha G$$

With substitution

$$e\,(\lambda) = \exp\,(r\Lambda)$$

we get characteristic equation

$$r^2 + 2\zeta r + 1 - \alpha = 0$$

with

$$r_{1,2} = -\zeta \pm \sqrt{\zeta^2 - 1 + \alpha}$$

Now if,

$$\alpha + \zeta^2 > 1$$

then the solution reads

$$e\,(\lambda) = C_1 \exp[(-\zeta + \sqrt{\alpha + \zeta^2 - 1})\lambda] + c_2 \exp[(-\zeta - \sqrt{\alpha + \zeta^2 - 1})\lambda]$$
$$+\frac{\alpha G}{1-\alpha}$$

Since

$$e\,(0) = 0, \quad c_1 + c_2 + \frac{\alpha G}{1-\alpha} = 0$$

and since $e(0) = 0$

$$c_1(-\zeta + \sqrt{\alpha^2 + \zeta^2 - 1}) + c_2(-\zeta^2 - \sqrt{\alpha^2 + \zeta^2 - 1}) = 0$$

We get

$$c_1 = \frac{\alpha G}{2(1-\alpha)}\left(1 + \frac{\zeta}{\sqrt{\alpha^2+\zeta^2-1}}\right)$$

$$c_2 = \frac{\alpha G}{2(1-\alpha)}\left(1 + \frac{\zeta}{\sqrt{\alpha^2+\zeta^2-1}}\right)$$

Finally

$$e(\lambda) = \frac{\alpha G}{(1-\alpha)}\frac{1}{2}\left(1 + \frac{\zeta}{\sqrt{\alpha^2+\zeta^2-1}}\right)\exp[(-\zeta+\sqrt{\alpha^2+\zeta^2-1})\lambda]$$
$$+\frac{1}{2}\left(1 - \frac{\zeta}{\sqrt{\alpha^2+\zeta^2-1}}\right)\exp[(-\zeta-\sqrt{\alpha^2+\zeta^2-1})\lambda]+1$$

The amplitude of the total displacement becomes

$$v(\lambda) = G + e(\lambda)$$
$$= G\left\{a + \frac{\alpha}{1-\alpha}\left\{\frac{1}{2}\left(1 - \frac{\zeta}{\sqrt{\alpha^2+\zeta^2-1}}\right)\exp[(-\zeta+\sqrt{\alpha^2+\zeta^2-1})\lambda]\right.\right.$$
$$\left.\left.+\frac{1}{2}\left(1 - \frac{\zeta}{\sqrt{\alpha^2+\zeta^2-1}}\right)\exp[(-\zeta-\sqrt{\alpha^2+\zeta^2-1})\lambda]\right\}\right.$$

Analogously we find expressions for $V(\lambda)$ for cases when $\zeta^2 + \alpha < 1$ and $\zeta^2 + \alpha = 1$. Further, probabilistic analysis, should follow steps underlined after equation (5.93).

PROBLEM 5.20

Consider the load-bearing capacity of an imperfect bar. As can be seen from the figure, a concentric load P produces a bending moment $M_z = -Pw$ and increases the displacement by an amount $w - w_1$. The differential equation of the column is, therefore,

$$EI_z \frac{d^2}{dx^2}(w - w_1) = -Pw \tag{5.124}$$

The bar is simply supported at its ends and has an initial imperfection

$$w_1 = g \sin\frac{\pi x}{l} \tag{5.125}$$

Equation (5.124) becomes, upon substitution of Eq. (5.125),

$$\frac{d^2w}{dx^2} + \frac{Pw}{EI_z} = -a\frac{\pi^2}{l^2}\sin\frac{\pi x}{l}$$

The solution of this equation is

$$w = C_1 \sin\left[\left(\frac{P}{EI_z}\right)^{1/2} x\right] + C_2 \cos\left[\left(\frac{P}{EI_z}\right)^{1/2} x\right] + \frac{a_1}{1 - P/P_c}\sin\frac{\pi x}{L} \tag{5.126}$$

where $P_c = \pi^2 EI/l^2$ is the classical or Euler buckling load. The boundary conditions are $w = 0$ at $x = 0, l$, and these yield

$$C_1 \sin\sqrt{\frac{P}{EI_z}}l = 0 \quad C_2 = 0$$

For $P < P_c$, both C_1 and C_2 must be zero, and the total deflection is represented by the last term in Eq. (5.126):

$$w = \frac{1}{1 - P/P_c} g \sin\frac{\pi x}{l} \tag{5.127}$$

We see that the total deflection becomes increasingly large, as $P \to P_c$. The normal stress in the bar are given

$$\sigma_x = -\frac{P}{A} - \frac{My}{I_z}, \quad M = -Pw$$

Thus the maximum compressive stress takes place at $x = 1/2$ and is given by

$$\sigma_{max} = \frac{P}{A}\left[1 + \frac{gA}{S}\frac{1}{1 - P/P_c}\right] \tag{5.128}$$

where S is the section modulus ($S = I_z/y_{max}$, y_{max} being the distance from the neutral axis to the point of maximum stress). Denote $P/A = \sigma_{av}$ and $P/P_c = \sigma_{av}/\sigma_c$. Equation (51.27) becomes then

$$\sigma_{max} = \sigma_{av}\left[1 + \frac{gA}{S}\frac{1}{1 - \sigma_{av}/\sigma_c}\right]$$

where $\sigma_c = \pi^2 E/(l/r)^2$, r being the radius of gyration of the cross-sectional area of the bar. The load P_L, for which σ_{max} equals the yield stress σ_y, is the limit load for which the column remains elastic. This load results in the average stress $\sigma_L = P_L/A$, and Eq. (5.127) becomes

$$\sigma_y = \lambda_L\left[1 + \frac{gA}{S}\frac{1}{1 - \sigma_L/\sigma_c}\right]$$

where

$$\left(\frac{\sigma_L}{\sigma_y}\right)^2 - \left[1+\frac{\sigma_c}{\sigma_y}\left(1+\frac{gA}{S}\right)\right]\frac{\sigma_L}{\sigma_y}+\frac{\sigma_c}{\sigma_y}=0 \tag{5.129}$$

Part (c) of the figure shows σ_L/σ_y as a function of the *slenderness ratio* l/r. Treating the initial imperfection amplitude G as a random variable with gamma distribution (Sec. 4.8),

$$f_G(g) = \frac{1}{\beta^{\alpha+1}\Gamma(\alpha+1)} g^{\alpha} e^{-g/\beta} U(g)$$

the average limit stress is also a random variable.

(a) Extract σ_L/σ_y explicitly from Eq. (5.128) as a function of g.
(b) Find $F_{\sum}(\sigma_L)$.
(c) Consider also the special cases of an exponential distribution and of a chi-square distribution with m degrees of freedom. Perform the numerical calculations.

SOLUTION 5.20

Note to lecturer: question 5.20 "b" should read: Find $F_{\sum_L(\sigma_L)}$ [instead of $F_{\sum(\sigma_L)}$].

(a) Equation (5.129) reads

$$\left(\frac{\sigma_L}{\sigma_y}\right)^2 - \left[1+\frac{\sigma_c}{\sigma_y}\left(1+\frac{gA}{S}\right)\right]\frac{\sigma_L}{\sigma_y}+\frac{\sigma_c}{\sigma_y}=0$$

which yields in

$$\frac{\sigma_L}{\sigma_y} = 1+\frac{\sigma_c}{\sigma_y}\left(1+\frac{gA}{S}\right) \pm \sqrt{\left[1+\frac{\sigma_c}{\sigma_y}\left(1+\frac{gA}{s}\right)\right]^2 - 4\frac{\sigma_c}{\sigma_y}}$$

b) In order to find $F_{\sum_L(\sigma_L)}$, we find

$$F_{\sum_L(\sigma_L)} = Prob\left(\sum L \le \sigma_L\right)$$

$$= Prob\left[\sigma_y + \sigma_c\left(1+\frac{A}{S}\right) \pm \sqrt{\left[\sigma_y+\sigma_c\left(1+\frac{A}{S}\right)\right]^2 - 4\sigma_c} \le \sigma_L\right]$$

$$= Prob\left(G \le g^*(\sigma_L)\right)$$

with

$$g^*(\sigma_L) = \left\{ \left[\frac{\left(\frac{\sigma_L}{\sigma_y}\right)^2 - \frac{\sigma_c}{\sigma_y}}{\sigma/\sigma_y} - 1 \right] \frac{\sigma_y}{\sigma_c} - 1 \right\} \frac{S}{A}$$

Therefore

$$F_{\sum}(\sigma_L) = F_G[g^*(\sigma_L)]$$

since G is gamma distributed, with

$$f_G(g) = \frac{1}{\beta^{\alpha+1}\Gamma(\alpha+1)} g^\alpha e^{-g/\beta} U(g)$$

the distrubtion function is for a integer:

$$f_G(g) = 1 - \exp\left(-\frac{g}{\beta}\right)\left[1 + \frac{g}{\beta} + \left(\frac{g}{\beta}\right)^2 + \cdots + \left(\frac{g}{\beta}\right)^\alpha\right]$$

[compare with Formula 7.68 and its proof, in section "Comment on Example 7.11"].

Cases of exponential or chi-square distribution of initial imperfections are considered in complete analogy.

PROBLEM 5.21

In problem 5–20, assume G has an exponential distribution. Find the probability of the maximum total displacement $w(l/2)$ taking on values in the interval $[0, 2E(G)]$. Investigate the behavior of this probability as P approaches the classical buckling load P_c.

SOLUTION 5.21

The displacement is given by Equation 5.127 (p. 167):

$$w(x) = \frac{1}{1 - P/P_c} g \sin\frac{\pi x}{l}$$

so that the maximum total deflection, is

$$W \equiv w\left(\frac{l}{2}\right) = \frac{1}{1 - P/P_c} G$$

If G is exponentially distributed, then

$$F_G(g) = 1 - \exp\left[-\frac{g}{E(g)}\right]$$

Now the sought probability P^* becomes:

$$\begin{aligned} P^* &= Prob\,[W < 2E(G)] \\ &= Prob[\frac{1}{1 - P/P_c} \cdot G \leq 2E(G)] \\ &= Prob[G \leq 2E(G)(1 - P/P_c)] \\ &= F_G[2E(G)(1 - P/P_c)] \\ &= 1 - \exp[-2(1 - P/P_c] \end{aligned}$$

For $P = 0$,

$$P^* = 1 - \exp(-2)$$

whereas for $P \to P_c$

$$P^* = 0.$$

since the displacements grow without bound, and the probability that they are smaller than $2E(G)$ (or bounded), vanishes.

PROBLEM 5.22

Consider now another important case of an initially perfect simply supported bar under eccentric load with eccentricity ε, as shown in part (a) of the accompanying figure. The differential equation for the bar deflection reads

$$EI_z \frac{d^2 w}{dx^2} + Pw = 0 \tag{5.130}$$

The solution is

$$w = C_1 \sin\sqrt{\frac{P}{EI_z}}x + C_2 \cos\sqrt{\frac{P}{EI_z}}x$$

The integration constants are determined by the boundary conditions $w = \varepsilon$ at $x = \pm l/2$, so that

$$C_1 = 0 \quad C_2 = \varepsilon \sec\left[\left(\frac{P}{EI_z}\right)^{1/2}(\frac{l}{2})\right] = \varepsilon \sec\left[\frac{\pi}{2}\left(\frac{P}{P_c}\right)^{1/2}\right] \tag{5.131}$$

It is seen that the maximum deflection becomes increasingly large as P approaches P_c. The elastic limit is reached at the most stressed point when

$$\sigma_y = \frac{P}{A} + \frac{M_c}{I_z} \tag{5.132}$$

or when

$$\sigma_0 = \frac{P}{A} + \frac{P\varepsilon c}{I_z}\sec\left[\frac{\pi}{2}\left(\frac{P}{P_c}\right)^{1/2}\right] \tag{5.133}$$

This equation can be rewritten as

$$\sigma_y = \sigma_L\left\{1 + \frac{\varepsilon c}{r^2}\sec[\frac{l}{2r}\left(\frac{\sigma_L}{E}\right)^{1/2}]\right\} \tag{5.134}$$

where $\sigma_L = P_L/A$ is the "average" limit stress. Equation (5.134) is referred to as the *secant formula* and is usually plotted in the form of σ_L/σ_y versus l/r for a particular material (with σ_y and E) specified for different values of $\varepsilon c/r^2$, as shown in part (b) of the figure.

Treating the eccentricity as the random variable E, with $F_E(\varepsilon)$ given an exponentially distributed, the average limit stress also is a random variable. Find $F_\Sigma(\sigma_L)$.

SOLUTION 5.22

$$Prob\left(\sum < \sigma\right) = Prob\left(E < \frac{\frac{\sigma_y}{\sigma} - 1}{\frac{c}{r^2}\sec\left[\frac{l}{2r}\left(\frac{\sigma}{E}\right)^{\frac{1}{2}}\right]}\right).$$

Therefore

$$F_\Sigma(\sigma_L) = F_E\left(\frac{\frac{\sigma_y}{\sigma_L} - 1}{\frac{c}{r^2}\sec\left[\frac{l}{2r}\left(\frac{\sigma_L}{E}\right)^{\frac{1}{2}}\right]}\right)$$

PROBLEM 5.23

In his (now historic) doctoral thesis in 1945, and in his 1963 paper, Koiter analyzed a sufficiently long cylindrical shell with an axisymmetric initial imperfection, under axial load. He chose an initial imperfection function $w_0(x)$, coconfigurational with the axisymmetric buckling mode of a perfect cylindrical shell, as

$$w_0(x) = gh \sin \frac{\pi i_c x}{L}, \quad i_c = \frac{L}{\pi}\left(\frac{2c}{Rh}\right)^{1/2}, \quad c = [3\left(1-\nu^2\right)]^{1/2} \qquad (5.135)$$

where μ is the nondimensional initial imperfection magnitude; i_c, the number of half-waves at which the associated perfect shell buckles; L, the shell length; R, the shell radius; and h, the shell thickness. Using his own general nonlinear theory, he derived *inter alia* a relationship between the critical load and the initial imperfection magnitude:

$$(1-\lambda)^2 - \frac{3}{2}c\,|g|\,\lambda = 0 \qquad (5.136)$$

where $\lambda = P_{bif}/P_c$, $P_c = 2\pi Rh\sigma_c$, and $\sigma_c = Eh/Rc$, is the nondimensional buckling load, P_{bif}, the buckling load of an imperfect shell; P_c, the classical buckling load of a perfect shell; E, the modulus of elasticity; and ν, Poisson's ratio. The buckling load P_{bif}, was defined as that at which the axisymmetric fundamental equilibrium state bifurcates into a nonsymmetric one. The absolute value of g in Eq. (5.136) stands, since for a sufficiently long shell the sign of the imperfection is immaterial: Positive and negative initial imperfections with equal absolute values cause the same reduction on the buckling load.

Equation (5.136) yields the explicit buckling load-initial imperfection relationship:

$$\lambda = 1 + \frac{3}{4}|\xi| - \frac{1}{2}\left(6|\xi| + \frac{9}{4}|\xi|^2\right)^{1/2}$$

where $\xi = cg$.

(a) Assume X with possible values ξ to be a normally distributed random variable $N(\bar{\xi}, \sigma^2)$.
(b) Find the probability density function of $|X|$. (Consult Example 4.6.)
(c) Find the reliability of the shell at the load level $\bar{\lambda}$.
(d) Assume, after Roorda, $\bar{\xi} = 0.333 \times 10^{-3} R/h$ and $\sigma^2 = 10^{-3} R/h$, and find the stress level at which the system had a given reliability. Compare your result with Roorda's.

SOLUTION 5.23

a) The reliability

$$
\begin{aligned}
R &= Prob(\Lambda > \lambda) \\
&= Prob\left(c\,|G| \le \frac{2\,(1-\lambda)^2}{3\lambda}\right) \\
&= Prob\left(|X| \le \frac{2\,(1-\lambda)^2}{3\lambda}\right)
\end{aligned}
$$

with $X = cG$. Now, if X is $N(\bar{\xi}, \sigma^2)$ we get for $Y = |X|$

$$F_Y(y) = \frac{1}{\sigma\sqrt{2\pi}} \exp\left[-\frac{\left(y-\bar{\xi}\right)}{2\sigma^2}\right] + \exp\left[-\frac{\left(y+\bar{\xi}\right)^2}{2\sigma^2}\right]$$

and

$$F_Y(y) = \mathrm{erf}\left(\frac{y-\bar{\xi}}{\sigma}\right) + \mathrm{erf}\left(\frac{y+\bar{\xi}}{\sigma}\right)$$

(see Equations (4.56) and 4.66).

Now

$$F_G(g) = f_X(\psi(g))\left|\frac{d\psi}{dg}\right|$$

with

$$x = \psi(g) = g\frac{h}{R}$$

we get

$$f_G(g) = \frac{h}{R} f_X(\psi(g))$$

or

$$f_G(y) = \begin{cases} \frac{h}{R}\frac{1}{10^{-3}-10^{-4}} & \text{for } 10^{-4} < g\frac{h}{R} < 10^{-3} \\ 0, & \text{otherwise} \end{cases},$$

or say for $R/h = 900$, we get

$$f_G(g) = \begin{cases} \dfrac{1}{900} \dfrac{1}{10^{-3} - 10^{-4}}, & \text{for } 10^{-4} < g\dfrac{h}{R} < 10^{-3} \\ 0, & \text{otherwise} \end{cases}$$

Finally

$$f_G(g) = \begin{cases} \dfrac{1}{0.9 - 0.09} = \dfrac{1}{0.81} & 0.09 \leq g \leq 0.9 \\ 0, & \text{otherwise} \end{cases},$$

Likewise,

$$R(\lambda) = Prob\left(Y = |X| < \frac{2(1-\lambda)^2}{3\lambda}\right)$$

$$= \text{erf}\left[\frac{2(1-\lambda)^2}{3\lambda\sigma} - \frac{\bar{\xi}}{\sigma}\right] + \text{erf}\left[\frac{2(1-\lambda)^2}{3\lambda \cdot 10^{-3} R/h} + 0.333\right]$$

In order to gain more insight, we solve this problem also for uniformly distributed G. Say

$$G = U \cdot R/h$$

where U is an uniformly distributed random variable between 10^{-3} and 10^{-4}.

Then

$$f_U(u) = \begin{cases} \dfrac{1}{10^{-3} - 10^{-4}}, & \text{for } 10^{-4} \leq u \leq 10^{-3} \\ 0, & \text{otherwise} \end{cases}$$

$$f_G(g) = \begin{cases} 0, & \text{for } g \leq 0.09 \\ \dfrac{g - 0.09}{0.81}, & \text{for } 0.09 \leq g \leq 0.9 \\ 1, & g > 0.9 \end{cases}$$

Now

$$R(\lambda) = Prob\,(\Lambda \geq \lambda) = Prob\left(|G| \leq \frac{2(1-\lambda)^2}{3c\lambda}\right)$$

But G gets only positive values, therefore

$$R(\lambda) = f_G\left[\frac{2(1-\lambda)^2}{3c\lambda}\right]$$

hence

$$R(\lambda) = \begin{cases} 0, & \text{for} \dfrac{2(1-\lambda)^2}{3c\lambda} \leq 0.09 \\ \dfrac{2(1-\lambda)^2/3c\lambda - 0.09}{0.81}, & \text{for } 0.09 \leq \dfrac{2(1-\lambda)^2}{3c\lambda} \leq 0.9 \\ 1, & \dfrac{2(1-\lambda)^2}{3c\lambda} > 0.9 \end{cases}$$

If $R_{reg} = r = 0.99$ we get

$$R(\lambda) = \frac{2(1-\lambda)^2/3c\lambda - 0.09}{0.81} = 0.99$$

This yields in

$$\frac{2(1-\lambda)^2}{3c\lambda} = 0.8919$$

or since

$$\lambda = 1 + \frac{3}{4}cg - \frac{1}{2}\sqrt{6cg + \frac{g}{4}c^2g^2}$$

we get

$$\lambda_{design} = 0.2527$$

PROBLEM 5.24

(a) Koiter (1945) also analyzed the imperfection sensitivity of a shell with nonaxisymmetric, periodic imperfections:

$$w_0(x) = gh\left(\cos\frac{i_c\pi x}{L} + 4\cos\frac{i_c\pi x}{2L}\cos\frac{i_c\pi y}{2L}\right) \tag{5.137}$$

(where y is the circumferential coordinate, the remaining notation as in the preceding problem) to arrive, instead of Eq. (5.136), at the equation

$$(1-\lambda)^2 + 6cg\lambda = 0 \tag{5.138}$$

for the nondimensional buckling load $\lambda = P_{lim}/P_c$, where P_{lim} is the limit load (as in Sec. 5.5). For the imperfection function (5.138), the limit load exists only at negative values of the imperfection parameter g. For positive g, the origin of the coordinate system may be shifted, and since the shell is sufficiently long, the analysis would be unaffected except that the sign of g would change to yield

$$(1-\lambda)^2 - 6cg\lambda = 0 \tag{5.139}$$

Combining Eqs. (5.138) and (5.139), we arrive at the final equation

$$(1-\lambda)^2 - 6c|g|\lambda = 0$$

Perform calculations as in Prob. 5.23. Compare the reliabilities of shells with axisymmetric and nonaxisymmetric imperfections.

(b) Sometimes the imperfection is represented by a local dimple extending over a small region of the shell. A more or less localized imperfection may be represented in the form

$$w_0(x) = gh\left(\cos\frac{i_c\pi x}{L} + 4\cos\frac{i_c\pi x}{2L}\cos\frac{i_c\pi y}{2L}\right) \tag{5.140}$$

$$\times \exp\left[-\frac{1}{2}\frac{\mu^2}{R^2}\left(x^2+y^2\right)\right] \tag{5.141}$$

which is the function in Eq. (5.137) multiplied by an exponentially decaying function. For example, at a distance $x = (4\pi/i_c)R$ or $y = (4\pi/i_c)R$, a complete wavelength of periodic part in Eq. (5.140), the exponential factor reduces to $\exp(-8\pi^2\mu^2/i_c^2)$. At first approximation, the term μ^2/i_c^2 may be neglected with respect to unity. Koiter's analysis (1978) yields then

$$(1-\lambda)^2 = -4cg\lambda \tag{5.142}$$

Assume again that $X = cG$, with possible values $\xi = cg$, is a normally distributed random variable $N(0, \sigma^2)$, and find the reliability of the shell. Are the localized imperfections as harmful as the periodic ones?

SOLUTION 5.24

Under new circumstances, for nonsymmetric imperfections,

$$\begin{aligned} R &= Prob(\Lambda \geq \lambda) \\ &= Prob\left(c\,|G| < \frac{(1-\lambda)^2}{6\lambda}\right) \\ &= Prob\left(|X| < \frac{(1-\lambda)^2}{6\lambda}\right) \end{aligned}$$

versus reliability for the case of the axisymmetric imperfections

$$R^5 = Prob\left(|X| \leq \frac{2}{3}\frac{(1-\lambda)^2}{\lambda}\right)$$

If parameter X is identically distributed in both cases, then at the same level of the loading λ

$$\frac{2}{3}\frac{(1-\lambda)^2}{\lambda} > \frac{1}{6}\frac{(1-\lambda)^2}{\lambda}$$

and for axisymmetric case the integration will be performed in a wider range. Therefore, the reliability of nonaxisymmetrically imperfect shell is not higher than that of the shell with axisymmetric imperfections.

We first assume that G takes on only positive values. Then

$$R = Prob\,(\Lambda \geq \lambda) = Prob\left[cG \leq \frac{(1-\lambda)^2}{4\lambda}\right] = F_{cG}\left[\frac{(1-\lambda)^2}{4\lambda}\right]$$

Let us compare the reliability of the shell with local imperfections, with that of the shell with periodic, nonaxisymmetric general, imperfections. In this case

$$R = Prob\left[cG < \frac{(1-\lambda)^2}{6\lambda}\right] = F_{cG}\left[\frac{(1-\lambda)^2}{6\lambda}\right]$$

Now, since the distribution function is a nondecreasing function of its argument,

$$F_{cG}\left[\frac{(1-\lambda)^2}{4\lambda}\right] > F_{cG}\left[\frac{(1-\lambda)^2}{6\lambda}\right]$$

implying that the reliability of the shell with local imperfections is not less than that of the shell with periodic nonsymmetric imperfections.

chapter 6

Two or More Random Variables

PROBLEM 6.1

The random variables X and Y are said to have a uniform distribution in $x^2 + y^2 \leq R^2$ if

$$f_{XY}(x, y) = \begin{matrix} \dfrac{1}{\pi R^2}, & x^2 + y^2 \leq R^2 \\ 0, & \text{otherwise} \end{matrix}$$

Verify that

$$f_X(x) = \begin{matrix} \dfrac{2}{\pi R}\sqrt{1 - \left(\dfrac{x}{R}\right)^2}, & |x| \leq R \\ 0, & \text{otherwise} \end{matrix}$$

that is, that X is not uniformly distributed. Find $f_Y(y)$.

SOLUTION 6.1

According to Equation (6.17) we have

$$f_X(x) = \int_{-\infty}^{\infty} f_{XY}(x, y)dy$$

where

$$f_{Xy}(x, y) = \begin{cases} \dfrac{1}{\pi R^2}, & \text{for } X^2 + y^2 \leq R^2 \\ 0, & \text{otherwise} \end{cases}$$

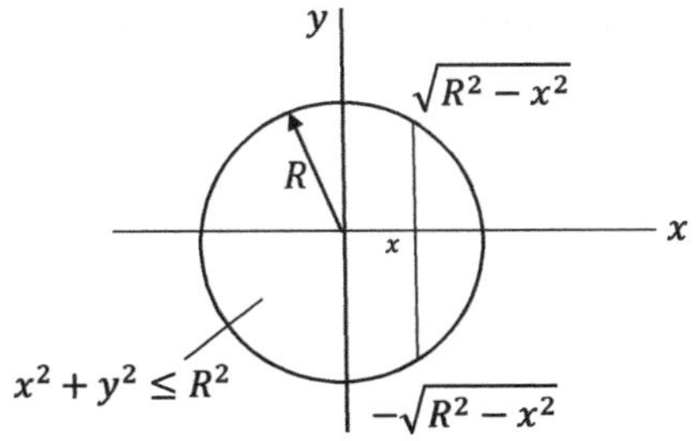

Therefore

$$f_X(x) = \int_{-\sqrt{R^2-x^2}}^{\sqrt{R^2-x^2}} \frac{1}{\pi R^2} dy = \frac{2\sqrt{R^2 - x^2}}{\pi R^2}, \quad \text{for } |x| \leq R$$

and zero otherwise

Analogically

$$f_Y(y) = \begin{cases} \frac{2}{\pi R^2}\sqrt{R^2 - y^2} & \text{for } (y) \leq R \\ 0, & \text{otherwise} \end{cases}$$

that is, neither X nor Y is uniformly distributed, despite the fact that jointly both of them are uniformly distributed on a circle.

PROBLEM 6.2

Find the conditional distribution function of X under the hypothesis that $Y \leq y$, $F_X(x|Y \leq y)$. Determine $f_X(x|Y \leq y)$.

SOLUTION 6.2

From formula (6.21) we obtain, putting $y_1 \to -\infty$, $y_2 = y$:

$$F_X(x \mid Y \leq y) = \frac{F_{XY}(x, y) - F_{XY}(x, -\infty)}{F_Y(y) - F_Y(-\infty)}$$

but

$$F_Y(-\infty) = 0, \; F_{XY}(x, -\infty) = 0$$

and

$$F_X(x \mid Y \leq y) = \frac{F_{XY}(x, y)}{F_Y(y)}$$

now, for the conditional probability density function we obtain

$$f_X(x \mid Y \leq y) = \frac{\partial}{\partial x} F_X(x \mid Y \leq y) = \frac{1}{F_Y\,(y)}(x, y)$$

or in terms of density functions,

$$F_X(x \mid Y \leq y) = \frac{\int_{-\infty}^{y} f_{XY}\,(x, n)\, dn}{\int_{-\infty}^{\infty} \int_{-\infty}^{y} f_{XY}\,(\xi, n)\, d\xi dn}$$

PROBLEM 6.3

Find the conditional distribution function $F_X(x|X \leq a,\ Y \leq b)$, provided that $F_{XY}(a, b) > 0$.

SOLUTION 6.3

In Equation (3.51) we defined the conditional distribution function of a random variable X under condition B, by

$$F_X\,(x \mid B) = P(X < x \mid B) = \frac{P(X \leq x, B)}{P(B)}$$

provided $P(B) \neq 0$.

Now we consider the case

$$B = \{X \leq a,\ Y \leq b\}$$

for which,

$$F_X\,(x \mid X \leq a,\ Y \leq b) = \frac{P\{X \leq a, Y \leq b, X \leq x\}}{P(X \leq a, Y \leq b)}$$

$$= \begin{cases} 1, X \geq a \\ F_{XY}(x, b) \\ F_{XY}(a, b) \end{cases}$$

provided $F_{XY}(ab) \neq 0$.

PROBLEM 6.4

Find the expression for the conditional distribution function $F_{XY}(x, y|a < X \leq b)$.

SOLUTION 6.4

According to Equation (3.51)

$$F_X(x \mid B) = P(X < x \mid B) = \frac{P(X \leq x, B)}{P(B)}$$

provided $P(B) \neq 0$.

Now we set

$$B = \{a \leq X \leq b$$

whereby

$$F_{XY}(x, y \mid a \leq X \leq b) = \frac{P(X \leq x, Y \leq y, a \leq X < b)}{P(a < X \leq b)}$$

$$= \begin{cases} \dfrac{F_{XY}(b, y) - F_{XY}(a, y)}{F_X(b) - F_X(a)} & \text{for } x > b \\ \dfrac{F_{XY}(x, y) - F_{XY}(a, y)}{F_X(b) - F_X(a)} & \text{for } a < x \leq b \\ 0 & \text{for } x \leq a \end{cases}$$

The appropriate density equals

$$f_{XY}(x, y \mid a \leq X \leq b) = \frac{f_{XY}(x, y)}{F_X(b) - F_X(a)} \quad \text{for } a \leq x \leq b$$

and vanishes for $x \leq a$ or $x \geq b$.

PROBLEM 6.5

Derive the formulas

$$f_X(x) = \int_{-\infty}^{\infty} f_X(x \mid y) f_Y(y) dy \quad \text{and}$$

$$f_Y(y) = \int_{-\infty}^{\infty} f_Y(y \mid x) f_X(x) dx$$

SOLUTION 6.5

According to equations (6.17) we have

$$f_X(x) = \int_{-\infty}^{\infty} f_{XY}(x, y) dy$$

$$f_Y(y) = \int_{-\infty}^{\infty} f_{XY}(x, y) dx$$

Now, in view of Equations (6.26) and (6.27),

$$f_{XY}(x, y) = f_X(x|y) f_Y(y)$$

$$f_{XY}(x, y) = f_Y(y|x) f_X(x)$$

Their substitution in the above formula yields the sought results.

PROBLEM 6.6

Show that the random variables defined in Prob. 6.1. are dependent.

SOLUTION 6.6

Due to Equations (6.26) and (6.27), we have

$$f_X(x \mid y) = \frac{f_{XY}(x, y)}{f_Y(y)}, \quad f_Y(y \mid x) = \frac{f_{XY}(x, y)}{f_X(x)}$$

substituting the results of Problem 6.1, we find

$$f_X\,(x \mid y) = \frac{\frac{1}{\pi R^2}\langle R^2 - (x^2 + y^2)\rangle}{\frac{2}{\pi R}\sqrt{1 - \left(\frac{1}{R}\right)^2}\,\langle R - |y|\rangle}$$

where

$$\langle R^2 - (x^2 + y^2)\rangle = \begin{cases} 1, & \text{if } R^2 \geq (x^2 + y^2) \\ 0, & \text{otherwise} \end{cases}$$

$$\langle R - |y|\rangle = \begin{cases} 1, & R \geq |y| \\ 0, & \text{otherwise} \end{cases}$$

The result is

$$f_X(x \mid y) = \begin{cases} \dfrac{1}{2R\sqrt{1 - (y/R)^2}} & \text{for } |X| < R\sqrt{1 - (y/b)^2} \\ 0, & \text{otherwise} \end{cases}$$

Analogically

$$f_Y(y \mid x) = \begin{cases} \dfrac{1}{2R\sqrt{1 - (x|R)^2}} & \text{for } |y| < R\sqrt{1 - (x/R)^2} \\ 0, & \text{otherwise} \end{cases}$$

We see that in this case the conditional probability densities differ from their unconditional counterparts, i.e. X and Y are dependent.

This conclusion can be drawn immediately by observing that Equation (6.30)

$$f_{XY}(x, y) = f_X(x) + f_Y(y)$$

valid for independent random variables, is violated in the present case.

The dependence of X and Y can be deduced directly, if we consider the range of values X and Y can take on. Indeed, it is clear that when X equals 0, then Y can take on all values from $-R$ to $+R$ with equal probability; but when X equals R, Y can only take on the single value of zero.

Let us check, in addition, whether X and Y are correlated. Since, due to symmetry,

$$E(X) = E(Y) = 0,$$

we are left with

$$Cov(X, y) = \int_\Omega\int xyf_{XY}(xy)dxdy = \frac{1}{\pi R^2}\iint_\Omega xydxdy$$

with integration over the circle $X^2 + y^2 \leq R^2$.

To perform the integration, we divide the integration domain into four sections Ω_1, Ω_2, Ω_3, and Ω_4, corresponding to four coordinate angles. In Ω_1 and Ω_3 the integral is positive, whereas in Ω_2 and Ω_4 it is negative. The absolute values of the integrals with respect to these sections equal:

$$\iint_{\Omega_1} xydxdy + \int_{\Omega_3}\int xydxdy = -\left[\int_{\Omega_2}\int xydxdy + \iint_{\Omega_4} +ydxdy\right]$$

therefore

$$\int_\Omega\int xydxdy = 0$$

and X and Y are uncorrelated. Since X and Y are dependent, as we saw above, we deduce that uncorrelatedness does not imply independence. (Re-read the discussion which follows Equation (6.71)).

Another example of this kind is given below.

Example

Let

$$X = \cos\phi, \quad Y = \sin\Phi$$

where Φ is uniformly distributed in the interval $(-\pi, \pi)$. We use interested in finding $Cov(X, Y)$.

SOLUTION

Since $E[\Phi] = 0$, then

$$E(X) = 0, \ E(Y) = 0.$$

Now, hence

$$Cov(XY) = E(XY) = E(cov\Phi \sin \Phi) = \frac{1}{2}E[\sin 2\Phi]$$
$$= \frac{1}{2}\int_{-\pi}^{\pi} \phi \cdot \sin 2\phi = 0$$

as an integral of an even function. This shows again, that despite the functional connection between X and $(X^2 + Y^2 = 1)$ the correlation between them vanishes.

PROBLEM 6.7

Find $E(X)$ and $E(Y)$ for the random variables in Prob. 6.1.

SOLUTION 6.7

Both mathematical expectations vanish

$$E(X) = 0, \; E(Y) = 0$$

since $f_{XY}(xy)$ is an even function with respect to both x and y. This can be verified by direct calculation — indeed

$$E(X) = \int_{-\infty}^{\infty} x f_Y(x) dx = \int_{-R}^{R} x \cdot \frac{2\sqrt{R^2 - x^2}}{\pi R^2} dx \equiv 0.$$

PROBLEM 6.8

X_1 and X_2 are independent standard normal random variables. Find $f_{Y_1 Y_2}(y_1, y_2)$, where $Y_1 = X_1 + X_2$, $Y_2 = X_1/X_2$. Verify that Y_2 has a Cauchy density.

SOLUTION 6.8

$$Y_1 = X_1 + X_2$$
$$Y_2 = X_1/X_2$$

from here

$$Y_1 = X_1 + \frac{X_1}{Y_2} = X_1\left(1 + \frac{1}{Y_2}\right) \to X_1 = \frac{Y_1}{1 + 1/Y_2} = \frac{Y_1 Y_2}{1 + Y_2}$$

$$X_2 = Y_1 - \frac{Y_1 Y_2}{1 + Y_2} = \frac{Y_1 Y_2 - Y_1 Y_2 - Y_1}{1 + Y_2} = \frac{Y_1}{1 + Y_2}$$

hence

$$x_1 = h_1(y_1, y_2) = \frac{y_1 y_2}{1+y_2}$$

$$x_2 = h_2(y_1, y_2) = \frac{y_1}{1+y_2}$$

The Jacobian of this transformation is (See page 225)

$$J = \begin{vmatrix} \frac{\partial h_1}{\partial y_1} & \frac{\partial h_1}{\partial y_2} \\ \frac{\partial h_2}{\partial y_1} & \frac{\partial h_2}{\partial y_2} \end{vmatrix} = \begin{vmatrix} \frac{y_2}{1+y_2} & y_1 \frac{(1+y_2)-y_2}{(1+y_2)^2} \\ \frac{1}{1+y_2} & \frac{y_1}{(1+y_2)^2} \end{vmatrix}$$

$$= -\frac{y_1 y_2 - y_1}{(1+y_2)^3} = -\frac{y_1(1+y_2)}{(1+y_2)^2} = -\frac{y_1}{(1+y_2)^2}$$

Equation (6.123) then yields

$$f_{\{Y\}}(y_1 \cdot y_2) = \left| \frac{\partial(h_1, h_2)}{\partial(y_1, y_2)} \right| f_{\{X\}} \left(\frac{y_1 y_2}{1+y_2}, \frac{y_1}{1+y_2} \right)$$

$$= \frac{|y_1|}{(1+y_2)^2} \frac{1}{2\pi} \exp \left\{ -\frac{1}{2} \left[\frac{(y_1 y_2)^2}{(1+y_2)^2} + \frac{y_1^2}{(1+y_2)^2} \right] \right\}$$

$$= \frac{1}{2\pi} \frac{|y_1|}{(1+y_2)^2} \exp \left[-\frac{1}{2} \frac{(1+y_2)^2 y_1^2}{(1+y_2)^2} \right]$$

To find the marginal density of Y_2 we must integrate with respect to y_1 yielding

$$f_{Y_2}(y_2) = \int_{-\infty}^{\infty} f_{Y_1 Y_2}(y_1, y_2) dy_1$$

$$= \frac{1}{2\pi} \frac{1}{(1+y_2)^2} \int_{-\infty}^{\infty} |y_1| \exp \left\{ -\frac{1}{2} \frac{(1+y_2)^2 y_1^2}{(1+y_2)^2} \right\} dy_1$$

We denote

$$z = \frac{1}{2} \frac{(1+y_2)^2 y_1^2}{(1+y_2)^2}$$

$$dz = \frac{(1+y_2)^2}{(1+y_2)^2} y_1 dy_1$$

which yields

$$f_{Y_2}(y_2) = \frac{1}{2\pi} \frac{1}{(1+y_2)^2} \frac{(1+y_2)^2}{1+y_2^2} \cdot 2 \cdot \int_0^{\infty} e^{-zdz}$$

$$= \frac{1}{\pi} \frac{1}{1+y_2^2}$$

implying that the ratio of two independent standard normal variables has a Cauchy density.

PROBLEM 6.9

$X_1, X_2, \ldots, X_n$ are independent random variables. Find the correlation coefficient r_{YZ}, where $Y = \sum_{j=1}^{n} \alpha_j X_j$, and $Z = \sum_{j=1}^{n} \beta_j X_j$, and α_j and β_j are given constants.

SOLUTION 6.9

$$Y = \sum_{j=1}^{n} \alpha_j X_j, \quad Z = \sum_{j=1}^{n} \beta_j X_j$$

The correlation coefficient is defined by Equation (6.68) and for our case

$$r_{YZ} = \frac{Cov(Y, Z)}{\sqrt{Var(Y)Var(Z)}}$$

Now

$$E(Y) = E\left(\sum_{j=1}^{n} \alpha_j X_j\right) = \sum_{j=1}^{n} \alpha_j E(X_j)$$

$$E(Z) = E\left(\sum_{j=1}^{n} \beta_j X_j\right) = \sum_{j=1}^{n} \beta_j E(X_j)$$

and via Equation (6.67)

$$Cov(Y, Z) = E(YZ) - E(Y)E(Z)$$

We calculate the first term, bearing in mind that j and k are dummy indices:

$$E(YZ) = E\left(\sum_{j=1}^{n} \alpha_j X_j \sum_{k=1}^{n} \beta_k X_k\right) = E\left(\sum_{j=1}^{n}\sum_{k=1}^{n} \alpha_j \beta_k X_j X_k\right)$$

$$= \sum_{j=1}^{n}\sum_{k=1}^{n} \alpha_j \beta_k E(X_j X_k)$$

But, since X_j and X_k are independent, they are also uncorrelated, meaning that

$$E(X_j X_k) = \begin{cases} E(X_j^2), & \text{for } j = k \\ 0, & \text{for } j \neq k \end{cases}$$

therefore

$$E(YZ) = \sum_{j=1}^{n} \alpha_j \beta_j E(X_j^2)$$

The variances equal

$$Var(Y) = E(Y^2) - [E(Y)]^2$$

$$E(Y^2) = E\left(\sum_{j=1}^{n} \alpha_j X_j \sum_{k=1}^{n} \alpha_k X_k\right) = \sum_{j=1}^{n} \sum_{j=k}^{n} \alpha_j \alpha_k E(X_j X_k)$$

and due to the uncorrelatedness of X_j and X_k

$$E(Y^2) = \sum_{j=1}^{n} \alpha_j^2 E(X_j^2)$$

So

$$Var(Y) = \sum_{j=1}^{n} \alpha_j^2 E\left(X_j^2\right) - \left[\sum_{j=1}^{n} \alpha_j E\left(X_j\right)\right]^2$$

$$Var(Z) = \sum_{j=1}^{n} \beta_j^2 E\left(X_j^2\right) - \left[\sum_{j=1}^{n} \beta_j E\left(X_j\right)\right]^2$$

Finally

$$r_{YZ} = \frac{\sum_{j=1}^{n} \alpha_j \beta_j E(X_j^2) - \left[\sum_{j=1}^{n} \alpha_j E(X_j)\left[\sum_{j=1}^{n} \beta_j E(X_j)\right]\right.}{\sqrt{\left\{\begin{array}{l} \left.\sum_{j=1}^{n} \alpha_j^2 E(X_j^2) - \left[\sum_{j=1}^{n} \alpha_j E(X_j)\right]^2\right\} \\ \times \sum_{j=1}^{n} \beta_j^2 E(X_j^2) - \left[\sum_{j=1}^{\infty} \beta_j E(X_j)\right]^2 \end{array}\right\}}}$$

PROBLEM 6.10

X_1 and X_2 are jointly normal random variables with $a = b = 0$, $\sigma_1 = \sigma_2 = 1$, and correlation coefficient r. Find $E[\max(X_1, X_2)]$ and $E[\min(X_1, X_2)]$.

SOLUTION 6.10

Denote

$$Y = \max(X_1, X_2)$$

The in virtue of Equation (6.63), we have

$$f_Y(y) = \int_{-\infty}^{y} f_{X_1X_2}(x_1x_2)dx_2 + \int_{-\infty}^{y} f_{X_1X_2}(x_1, x_2)dx_1$$

and

$$E(Y) = \int_{-\infty}^{\infty} y\left[\int_{-\infty}^{y} f_{X_1X_2}(x_1x_2)dx_2 + \int_{-\infty}^{y} f_{X_1X_2}(x_1x_2)dx_1\right]dy$$

where (see Equation 6.104)

$$f_{X_1X_2}(x_1, x_2) = \frac{1}{2\pi(1-r^2)^{1/2}} \exp\left[-\frac{1}{2(1-r^2)}\left(x_1^2 - 2rx_1x_2 + x_2^2\right)\right]$$

analogically

$$Z = \min(X_1, X_2)$$

with

$$f_Z(z) = f_{X_1}(z) + f_{X_2}(z) - \frac{d}{dz}F_{x_1x_2}(z, z)$$

(see last equation on page 190).

$$E(Z) = \int_{-\infty}^{\infty} z[f_{X_1}(z) + f_{X_2}(z) - \frac{d}{dz}F_{x_1x_2}(z, z)dz$$

$$= E(X_1) + E(X_2) - \int_{-\infty}^{\infty} z\frac{d}{dz}F_{X_1X_2}(z, z)dz$$

PROBLEM 6.11

X_1 and X_2 are jointly random variables. Find the joint distribution of

$$Y_1 = aX_1 + bX_2$$
$$Y_2 = cX_1 + dX_2$$

for constants a, b, c, and d satisfying $ad = bc$.

SOLUTION 6.11

Consider first the case

$$ad \neq bc$$

then the system

$$y_1 = ax_1 + bx_2$$
$$y_2 = ax_1 + dx_2$$

has a unique solution

$$x_1 = h_1(y_1, y_2) = a_1 y_1 + b_1 y_2$$

$$x_2 = h_2(y_1, y_2) = a_1 y_1 + d_1 y_2$$

for value of y_1 and y_2.

The Jacobian (see page 225) is

$$J = \begin{vmatrix} \frac{\partial h_1}{\partial y_1} & \frac{\partial h_1}{\partial y_2} \\ \frac{\partial h_2}{\partial y_1} & \frac{\partial h_2}{\partial y_2} \end{vmatrix} = \begin{vmatrix} a_1 & b_1 \\ c_1 & d_1 \end{vmatrix} = a_1 d_1 - b_1 c_1$$

and

$$f_{\{y\}}(y_1, y_2) = |J| f_X[h_1(y_1, y_2) h_2(y_1 y_2)]$$

$$= |a_1 d_1 - b_1 c_1| f_X(a_1 y_1 + b_1 y_2,\ c_1 y_1 + d_1 y_2)$$

in our case

$$a_1 = \frac{d}{ad - bc}, \quad b_1 = \frac{b}{ad - bc}$$

$$c_1 = -\frac{c}{ad - bc}, \quad d_1 = \frac{a}{ad - bc}$$

$$J = a_1 d_1 - b_1 c_1 = \frac{ad - bc}{(ad - bc)^2} = \frac{1}{ad - bc}$$

with

$$f_{\{y\}}(y_1, y_2) = \frac{1}{|ad - bc|} f_X \times \left(\frac{d}{ad - bc} y_1 - \frac{b}{ad - bc} y_2 - \frac{c}{ad - bc} y_1 + \frac{a}{ad - bc} y_2 \right)$$

Now if $ad = bc$, we denote

$$\frac{c}{a} = \frac{d}{b} = \alpha$$

so that

$$y_2 = c x_1 + d x_2 = \alpha(a x_1 + b x_2) = \alpha y_1$$

and $y_1 + y_2$ are dependent random variables. We first determine $f_{Y_1}(y_1)$. This is a simple generalization of Example 6.5 (see pp. 192–193), concerning the distribution of the sum of random variables, to the linear combination of the random variables. The domain D of the (x_1, x_2) — plane is determined by the equation $a_1 X_1 + b X_2 \leq y$, and is shown in the figure.

$$F_{Y_1}(y_1) = \int_{-\infty}^{\infty} dx_2 \int_{-\infty}^{\frac{y_1 - b x_2}{|a|}} f_X(x_1, x_2) dx_1$$

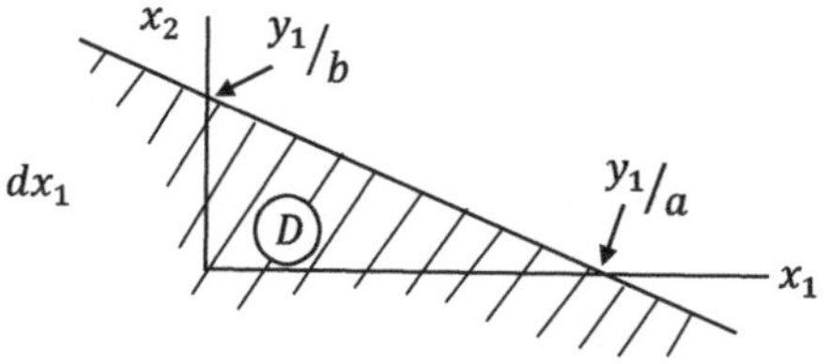

Differentially the equation with respect to y, we have

$$F_{Y_1}(y_1) = \frac{dF_{Y1}(y_1)}{dy_1}\frac{1}{|a|}\int_{-\infty}^{\infty} f_{\{X\}}\left(\frac{y_1 - bx_2}{a}, x_2\right) dx_2$$

Now we have to determine the density function Y_1 and $Y_2 = \alpha Y_1$. Let us find the conditional probability of Y_2 under the hypothesis $Y_1 = y_1$. For any values y_1 of the random variable Y_1, Y_2 has a single value $y_2 = \alpha y_1$. Therefore

$$F_{Y_1}(y_1) = \int_{-\infty}^{\infty} dx_2 \int_{-\infty}^{\frac{y_1 - bx_2}{|a|}} f_x(x_1, x_2)\, dx_1$$

Differentiating the equation with respect to y, we have

$$F_{Y_1}(y_1) = \frac{dF_{Y1}(y_1)}{dy_1}\frac{1}{|a|}\int_{-\infty}^{\infty} f_{\{X\}}\left(\frac{y_1 - bx_2}{a}, x_2\right) dx_2$$

$$f_{Y_2|Y_1}(y_2 \mid y_1) = \delta(y_2 - \alpha y_1)$$

since $[Y_2|Y_1\}$ is a casually distributed random variable (see Section 4.1) the joint density of Y_1 and Y_2 is:

$$f_{Y_1Y_2}(y_1, y_2) = f_{Y_1|Y_2}(y_2 \mid y_1)\, f_{Y_1}(y_1) = f_{Y_1}(y_1)\delta(y_2 - \alpha y_1)$$

which finally yields

$$f_{Y_1Y_2}(y_1, y_2) = \delta(y_2 - \alpha y_1)\frac{1}{|a|}\int_{-\infty}^{\infty} f_{\{X\}}\left(\frac{y_1 - bx_2}{a}, x_2\right) dx_2.$$

PROBLEM 6.12

$X_1, X_2, \ldots, X_n$ are independent identically distributed exponential random variables. Find $f_Y(y)$ of their sum $Y = X_1 + X_2 + \ldots + X_n$.

SOLUTION 6.12

As was shown in Equation (6.100), the characteristic function of the sum of independent random variables equals the product of the characteristic functions of the constituents

$$M_Y(\theta) = \prod_{j=1}^{n} M_{X_j}(\theta)$$

For X_j with identical distribution, i.e.

$$M_{X_j}(\theta) = M(\theta)$$

we have Equation (6.101)

$$M_Y(\theta) = [M(\theta)]^n$$

The characteristic function of the exponentially distributed random variable is

$$M_{X_j}(\theta) = \frac{a}{a - i\theta}$$

(see p. 75). Therefore:

$$M_Y(\theta) = \left(\frac{a}{a - i\theta}\right)^n$$

Using the inverse formula, Equation 3.41, we find

$$\begin{aligned} f_Y(y) &= \frac{1}{2\pi} \int_{-\infty}^{\infty} M_Y(\theta) e^{-i\theta y} d\theta \\ &= \frac{1}{2\pi} \int_{-\infty}^{\infty} \left(\frac{a}{a - i\theta}\right)^n e^{-i\theta y} d\theta \end{aligned}$$

Evaluation of this integral yields

$$f_Y(y) = \frac{a^n y^{n-1} e^{-ay}}{\Gamma(a)}$$

implying that Y has a gamma distribution (see p. 75) with parameter $\alpha + 1 = n$, $\beta = 1/a$.

This can also be deduced by comparing the characteristic functions $M_y(\theta)$ and of a gamma-distribution variable:

$M_Y(\theta) = \left(\frac{a}{a-i\theta}\right)^n$ versus that of (see p. 76)

$$\frac{1}{(1 - i\theta\beta)^{n+1}}$$

The distribution function of Y is given by

$$F_Y(y) = 1 - e^{-ay}\left\{1 + ay + \frac{(ay)^2}{2!} + \cdots + \frac{(ay)^{n-1}}{(n-1)!}\right\}, \ y \geq 0.$$

Consider the case $n = 2$ in more detail. This can be derived in accordance with Equation (6.52), indeed

$$f_Y(y) = \int_{-\infty}^{\infty} f_{X_2}(y - x_1) f_{X_1}(x_1) dx_1$$

Now

$$f_{X_1}(x) = f_{X_2}(x) = \begin{cases} ae^{-ax}, & \text{for } x > 0 \\ 0, & \text{otherwise} \end{cases}$$

Therefore, $f_Y(y) = 0$ for $y < 0$. for $y > 0\, Y$

$$f_Y(y) = \int_0^y f_{X_2}(y - x_1) f_{X_1}(x_1) dx_1$$

since $f_{X_2}(y - x_1) = 0$ for $y < x_1$. Hence

$$\begin{aligned} f_Y(y) &= \int_0^y ae^{-a(y-x_1)} ae^{-ax_1} dx_1 \\ &= a^2 e^{-ay} \int_0^y dx_1 = a^2 y e^{-ay} \end{aligned}$$

With

$$F_Y(y) = \int_0^y a^2 y e^{-ay} dy = 1 - e^{-ay}(1 + ay).$$

PROBLEM 6.13

$X_1, X_2, \ldots, X_n$ are independent identically distributed gamma-distributed random variables. Find $f_Y(y)$ of their sum $Y = X_1 + X_2 + \cdots + X_n$.

SOLUTION 6.13

As Equation (101) demonstrates, the characteristic function of the sum of indentically distributed random variables equals the nth power of the characteristic function of the single constituent of this sum, i.e.

$$M_Y(\theta) = [M(\theta)]^n$$

where, in virtue of the last equation on p. 76,

$$M(\theta) = \frac{1}{(1 - i\theta\beta)^{\alpha+1}}$$

so that

$$M_Y(\theta) = \frac{1}{(1 - i\theta\beta)^{n(\alpha+1)}}$$

Since the characteristic function uniquely describes the random variable, the latter characteristic function represents a gamma distributed random variable, with the density function obtainable directly form Equation (4.10) by replacing $\alpha + 1$ with $n(\alpha + 1)$:

$$f_Y(y) = \begin{cases} 0, & \text{for } y < 0 \\ \dfrac{1}{\beta^{n(\alpha+1)}\Gamma[n(\alpha + 1)]} x^{n(\alpha+1)-1} e^{-y/\beta}, & \text{for } y > 0 \end{cases}.$$

PROBLEM 6.14

$X_1, X_2, \ldots, X_n$ are independent and identically distributed normal random variables, $N(a, \sigma^2)$. Verify that the sample mean $\bar{X}_n = (1/n)\sum_{j=1}^{n} X_j$ is also a normal variable $N(a, \sigma^2/n)$.

SOLUTION 6.14

$$\bar{X}_n = \frac{1}{n}\sum_{j=1}^{n} X_j$$

$E(\bar{X}_n)$ and variance $V(\bar{X}_n)$ are readily found for any X_j (not necessarily normally distributed).

Indeed:

$$E(\bar{X}_n) = E\left(\frac{1}{n}\sum_{j=1}^{n} X_j\right) = \frac{1}{n^2}\sum_{j=1}^{n} E(X_j)$$

If the X_j's are identically distributed, then

$$E(X_j) = \text{const} = a$$

and

$$E(\bar{X}_n) = \frac{1}{n}\left[nE\left(X_j\right)\right] = E(X_j) = a$$

Now

$$Var(\bar{X}_n) = V\left(\frac{1}{n}\sum_{j=1}^{n} X_j\right) = \frac{1}{n^2}\sum_{j=1}^{n} Var(X_j)$$

and again, for identically distributed X_j,

$$Var(X_j) = \text{const} = \sigma^2$$

and

$$Var\left(\bar{X}_n\right) = \frac{1}{n^2}\cdot n\sigma^2 = \frac{\sigma^2}{n}$$

We prove that X_n is normally distributed, with the above mean and variance, i.e. $\bar{X}_n$ is $N(a, \sigma^2/n)$.

Let us find the characteristic function of $\bar{Y} = X_n$:

$$\begin{aligned} M_Y(\theta) &= E[\exp(i\theta Y)] \\ &= E\left\langle \exp\left\{ i\theta\left[\frac{1}{n}(X_1 + X_2 + \cdots + X_n)\right]\right\}\right\rangle \\ &\quad M_{\{X\}}\left(\frac{1}{n}\theta, \frac{1}{n}\theta, \ldots, \frac{1}{n}\theta\right) \\ &= \prod_{j=1}^{n} M_{X_j}\left(\frac{1}{n}\theta\right) \\ &= \sum_{j=1}^{n}\exp\left(ia\frac{\theta}{n} - \frac{\theta^2}{2}\cdot\frac{1}{n^2}\theta^2\right) \\ &= \exp\left(i\theta\frac{1}{n}\sum_{j=1}^{n} a - \frac{\theta^2}{n}\sum_{j=1}^{n}\frac{1}{n^2}\sigma^2\right) \\ &= \exp\left(i\theta a - \frac{\theta^2}{2}\frac{\theta^2}{n}\right) \end{aligned}$$

Comparing the last results and Equation (4.23), we established that the sample mean $\bar{X}_n$ is also normally distributed $N\left(a, \frac{\sigma^2}{n}\right)$ Q.E.D.

chapter 7

Reliability of Structures Described by Several Random Variables

PROBLEM 7.1

Both $X \equiv \sum$ and $Y \equiv \sum_{\text{allow}}$ have uniform distributions as per formula (7.13). Find the reliability in the following cases:

(a) $\gamma < \alpha$ and $\alpha \leq \delta \leq \beta$
(b) $\alpha \geq \gamma$, $\beta \geq \delta$

SOLUTION 7.1

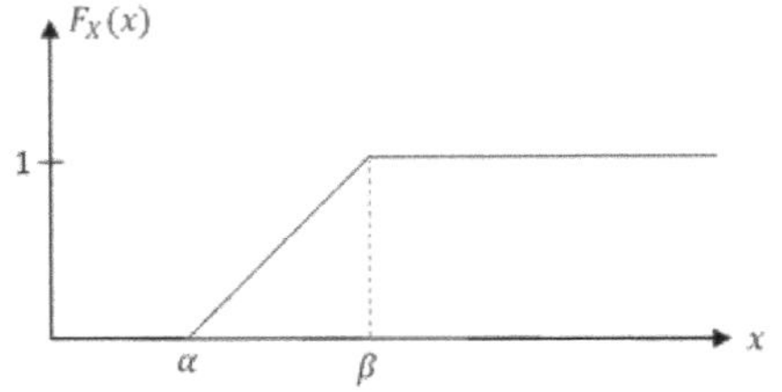

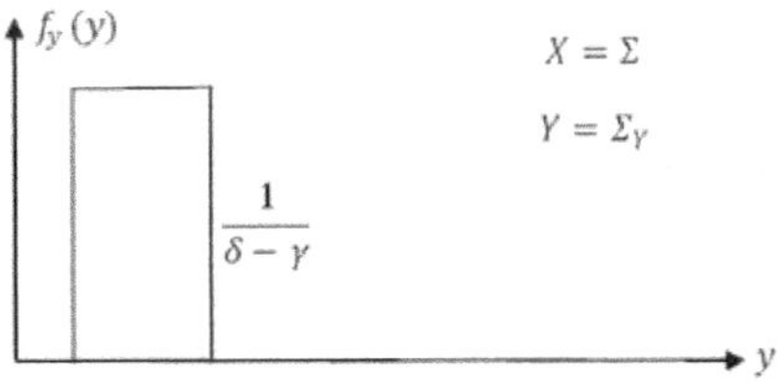

$$R = \frac{1}{\delta - \gamma}\int_{\alpha}^{\delta} \frac{y-\alpha}{\beta-\alpha} dy = \frac{1}{(\delta-\gamma)(\beta-\alpha)}\left[\frac{\delta^2-\alpha^2}{2} - \alpha(\delta-\alpha)\right]$$

$$= \frac{(\delta-\alpha)}{(\delta-\gamma)(\beta-\alpha)}\left[\frac{\delta+\alpha}{2} - \alpha\right] = \frac{(\delta-\alpha)^2}{2(\delta-\gamma)(\beta-\alpha)}$$

For say

$$\alpha = 12, \quad \beta = 20, \quad \gamma = 10, \quad \delta = 18$$

we get

$$R = \frac{(18-12)^2}{2(18-10)(20-12)} = \frac{36}{2 \cdot 8 \cdot 8} = \frac{9}{16}$$

(b) $\alpha > \gamma$, $\beta > \delta$

Here we have two cases, the first being $\alpha \leq \delta \leq \beta$ and this is as in above. The second probability is $\alpha > \delta$.

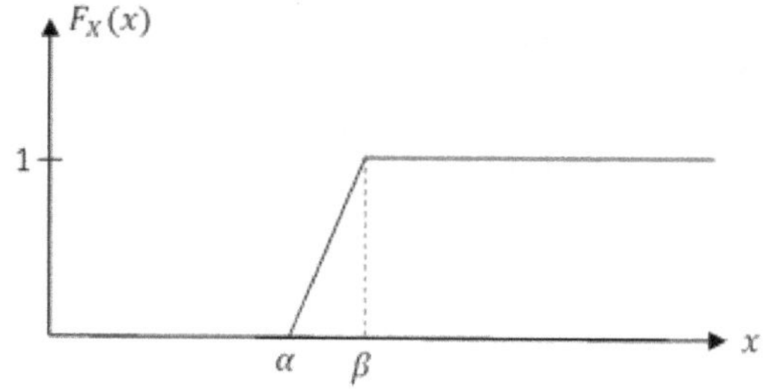

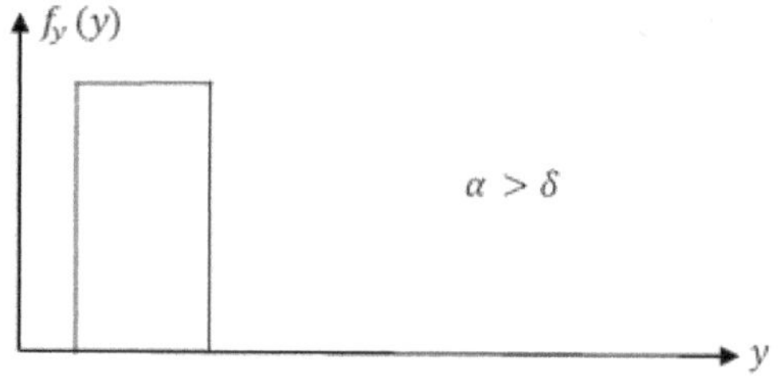

Then $R = 0$, since $minx_i = \alpha > maxy_i = \delta$.

PROBLEM 7.2

Both $X \equiv \sum$ and $Y \equiv \sum_{\text{allow}}$ have truncated normal distributions, as in Example 7.5. Find the probability density $f_Z(z)$ of the difference $Z = X - Y$ for $z < 0$.

SOLUTION 7.2

We have

$$f_X(x) = \phi(x)[U(x_2 - x) - U(x_1 - x)]$$

$$f_Y(y) = \psi(y)[U(y_2 - y) - U(y_1 - y)]$$

Where

$$\phi(x) = \frac{A}{\sigma_1\sqrt{2\pi}} \exp\left[-\frac{(x-m_1)^2}{2\sigma_1^2}\right]$$

$$\psi(y) = \frac{B}{\sigma_2\sqrt{2\pi}} \exp\left[-\frac{(y-m_1)^2}{2\sigma_2^2}\right]$$

The probability density for $Z = X - Y$ can be obtained through the formula 6.52 for the sum of two random variables; indeed we have a sum of X and $(-Y)$ so that this leads to

$$f_Z(z) = \int_{-\infty}^{\infty} f_X(z+y)\, f_Y(y)\, dy$$

$$= \int_{-\infty}^{\infty} \phi(z+y)\,\psi(y)\,[U(x_2 - z - y) - U(x_1 - z - y)]$$

$$\times\ [U(y_2 - y) - U(y_1 - y)]\, dy$$

PROBLEM 7.3

Given that the actual stress X and allowable stress Y are jointly normally distributed random variables with mathematical expectations m_1 and m_2, standard deviations σ_1 and σ_2 and correlation coefficient r.

Verify that

$$f_V(v) = \frac{\sqrt{1-r^2}}{\pi}\left(\frac{\sigma_1\sigma_2}{\sigma_1^2 - 2r\sigma_1\sigma_2 v + \sigma_2^2 v^2}\right)$$

$$\times \exp\left\{-\frac{1}{2(1-r^2)\,\sigma_1^2\sigma_2^2}\left[m_1^2\sigma_2^2 - 2rm_1m_2\sigma_1\sigma_2 + m_2^2\sigma_1^2\right]\right\}$$

where

$$V = \frac{X}{Y} \quad t = \frac{m_2\sigma_1^2 - rm_1\sigma_1\sigma_2 + m_1\sigma_2^2 v - rm_2\sigma_1\sigma_2 v}{\sigma_1\sigma_2\sqrt{(1-r^2)(\sigma_2^2 v - 2r\sigma_1\sigma_2 v + \sigma_1^2)}}$$

In the particular case $m_1 = m_2 = 0$,

$$f_V(v) = \frac{\sqrt{1-r^2}}{\pi}\left(\frac{\sigma_2/\sigma_1}{1 - 2r(\sigma_2/\sigma_1)v + (\sigma_2/\sigma_1)^2 v^2}\right)$$

With what density does the latter coincide for uncorrelated X and Y? Find the reliability of the structure.

SOLUTION 7.3

As is demonstrated in Example 6.7 (pages 194–195) the relevant integration domain is shown in Figure 6.9 (for $v = x/y$)

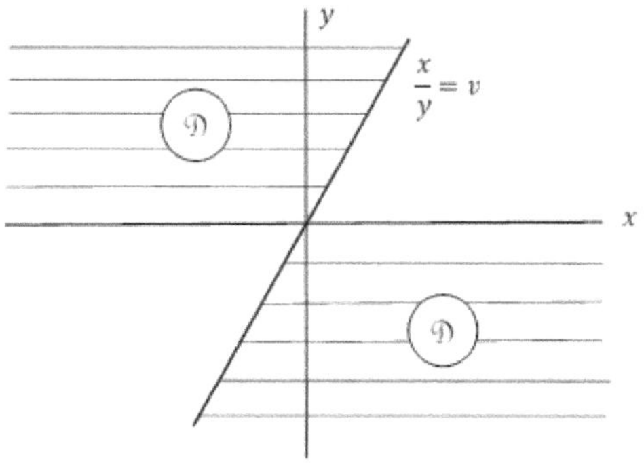

$$F_V(v) = \int_0^\infty dy \int_{-\infty}^{vy} f_{XY}(x, y)\, dx + \int_{-\infty}^{0} y f_{XY}(x, y)\, dx$$

To find the probability density $f_V(v)$ we differentiate

$$f_V(v) = \int_0^\infty y f_{XY}(vy, y)\, dy - \int_{-\infty}^{0} y f_{XY}(vy, y)\, dy$$

$$= \int_{-\infty}^{\infty} |y|\, f_{XY}(vy, y)\, dy$$

Substituting expression for $f_{xy}(xy)$, given by Equation 6.104

$$f_{XY}(xy) = \frac{1}{2\pi\sigma_1\sigma_2(1-r^2)^{1/2}} \exp\left\{-\frac{1}{2(1-r^2)}\left[\left(\frac{x-a}{\sigma_1}\right)^2 -2r\frac{x-a}{\sigma_1}\frac{y-b}{\sigma_2} + \left(\frac{y-b}{\sigma_2^1}\right)^2\right]\right\}$$

we get

$$f_V(v) = \frac{1}{2\pi\sigma_1\sigma_2(1-r^2)}\left[\int_0^\infty - \int_{-\infty}^0\right] y \exp\left\{-\frac{1}{2(1-r^2)}\left[\left(\frac{vy-a}{\sigma_1}\right)^2\right.\right.$$

$$\left.\left.-2r\frac{vy-a}{\sigma_1}\frac{y-b}{\sigma_2}+\left(\frac{y-b}{\sigma_2}\right)\right]\right\}dy$$

We get in square brackets

$$\left[\left(\frac{vy-a}{\sigma_1}\right)^2 - 2r\frac{vy-a}{\sigma_1}\frac{y-b}{\sigma_2}+\left(\frac{y-b}{\sigma_2}\right)^2\right]$$

$$= \frac{\sigma_2^2(vy-a)^2 - 2r\sigma_1\sigma_2(vy-a)(y-b)+\sigma_1^2(y-b)^2}{\sigma_1^2\sigma_2^2}$$

$$= \frac{y^2\left(\sigma_2^2v^2 - 2r\sigma_1\sigma_2 v+\sigma_1^2\right)+2y(-va\sigma_2^2+r\sigma_1\sigma_2 a+r\sigma_1\sigma_2 vb-\sigma_1^2 b}{\sigma_1^2\sigma_2^2}$$

$$+\frac{\sigma_2^2a^2-2r\sigma_1\sigma_2 ab+\sigma_1^2b^2}{\sigma_1^2\sigma_2^2}$$

consider now the particular case $m_1 = m_2 = 0$. Then the general expression for $f_V(v)$ becomes

$$f_V(v) = \frac{1}{2\pi\sigma_1\sigma_2\sqrt{1-r^2}}\left[\int_0^\infty - \int_{-\infty}^0\right] y$$

$$\times \exp\left\{-\frac{y^2}{2(1-r^2)}\left[\frac{\sigma_2^2v^2-2r\sigma_1\sigma_2 v+\sigma_1^2}{\sigma_1^2\sigma_2^2}\right]\right\}dy$$

$$= \frac{1}{\pi\sigma_1\sigma_2\sqrt{1-r^2}}\int_0^\infty y\exp\left\{-\frac{y^2}{2(1-r^2)}\frac{\sigma_2^2v^2-2r\sigma_1\sigma_2 v+\sigma_1^2}{\sigma_1^2\sigma_2^2}\right\}dy$$

We put now

$$z = \frac{y^2}{2(1-r^2)}\frac{\sigma_2^2v^2-2r\sigma_1\sigma_2 v+\sigma_1^2}{\sigma_1^2\sigma_2^2}$$

and

$$f_V(v) = \frac{\sigma_1\sigma_2\sqrt{1-r^2}}{\pi(\sigma_2^2 v^2 - 2r\sigma_1\sigma_2 v + \sigma_1^2)} \int_0^\infty e^{-z}dz = \frac{\sigma_1\sigma_2\sqrt{1-r^2}}{\pi(\sigma_2^2 v^2 - 2r\sigma_1\sigma_2 v + \sigma_1^2)},$$

Q.E.D.

For uncorrelated stress and strength, we get, putting $r = 0$:

$$f_V(v) = \frac{\sigma_1\sigma_2}{\pi\left(\sigma_1^2 + \sigma_2^2 v^2\right)} = \frac{\sigma_1\sigma_2^{-1}}{\pi(\sigma_1^2/\sigma_2^2 + v^2)}$$

which is the Cauchy density (see Example 3.8 on p. 51) with $a = \sigma_1/\sigma_2$. The distribution function is

$$F_V(v) = \int_{-\infty}^{v} \frac{\sigma_1/\sigma_2}{\pi}\frac{dv}{\sigma_1^2/\sigma_2^2 + v^2} = \frac{1}{2} + \frac{1}{\pi}\tan^{-1}\left(\frac{v}{\sigma_1/\sigma_2}\right)$$

The reliability is

$$F = F_V(1)$$

Hence

$$R = \frac{1}{2} + \frac{1}{\pi}\tan^{-1}\left(\frac{\sigma_2}{\sigma_1}\right)$$

PROBLEM 7.4

Assume in Example 7.7

$$f_{MT}(m,t) = \frac{1}{2\pi\sigma_M\sigma_T}\exp\left[-\frac{1}{2}\left(\frac{m^2}{\sigma_M^2} + \frac{t^2}{\sigma_T^2}\right)\right]$$

that is, M and T are independent normal variables with zero mean and with variances σ_M^2 and σ_T^2, respectively. (a) Verify that

$$f_M(m) = \frac{m}{\sigma_M\sigma_T}\exp\left[-\frac{m^2\left(\sigma_M^2 + \sigma_T^2\right)}{4\sigma_M^2\sigma_T^2}\right] I_0\left(\frac{m^2\left|\sigma_M^2 - \sigma_T^2\right|}{4\sigma_M^2\sigma_T^2}\right)$$

and find the reliability of the shaft. (b) Repeat for the case where $\sigma_M = \sigma_T$.

Note to lecturer:

In question (a) change $M \to Z(= M_{eq})$, $m \to z$ i.e.

$$f_Z(z) = \frac{z}{\sigma_M\sigma_T}\exp\left[-\frac{z^2\left(\sigma_m^2 + \sigma_T^2\right)}{4\sigma_m^2\sigma_T^2}\right] I_0\left(\frac{z^2\left|\sigma_m^2 - \sigma_T^2\right|}{4\sigma_m^2\sigma_T^2}\right)$$

SOLUTION 7.4

We begin with subcase (b), i.e., $\sigma_m = \sigma_T$, that is

$$f_{MT}(m,t) = \frac{1}{2\pi\sigma^2}\exp\left[-\frac{1}{2\sigma^2}\left(m^2+t^2\right)\right]$$

Now

$$F_Z(z) = \underset{\sqrt{m^2+t^2}\le z}{\int\int} f_{MT}(m,t)\,dmdt$$

with $m = r\cos\theta$, $t = r\sin\theta$ we get

$$F_Z(z) = \int_0^z\int_0^{2\pi}\frac{1}{2\pi\sigma^2}\exp\left(-\frac{r^2}{2\sigma^2}\right)(rdrd\theta)$$

$$= \frac{r}{2\pi\sigma^2}\int_0^z 2\pi r exp\left(-\frac{r^2}{2\sigma^2}\right)dr = 1-\exp\left(-\frac{z^2}{2\sigma^2}\right)$$

$$f_Z(z) = \frac{z}{\sigma^2}\exp\left(-\frac{z^2}{2\sigma^2}\right)U(Z)$$

Reliability is, in virtue of Equation (7.40)

$$R = F_Z\left(\frac{\pi}{4}\sigma_Y c^3\right) = 1-\exp\left(-\frac{1}{2\sigma^2}\cdot\frac{\pi^2\sigma_Y^2c^6}{16^1}\right);$$

and this coincides with Equation (7.51).

Consider now the case $\sigma_M \neq \sigma_T$. As we have shown in Example 7.7, the probability density $f_Z(z)$ is

$$f_Z(z) = \int_0^{2\pi} zf_{MT}(z\cos\theta,\ z\sin\theta)\,d\theta$$

(see Equation 7.43). Now, since M is $N(0,\sigma_M^2)$, and T is $N(0,\sigma_T^2)$,

$$f_{MT}(m,t) = \frac{1}{2\sigma_m^2\sigma_T^2}\exp\left(-\frac{m^2}{2\sigma_M^2}-\frac{t^2}{2\sigma_T^2}\right)$$

and

$$f_Z(z) = \frac{z}{2\pi\sigma_M\sigma_T}\int_0^{2\pi}\exp\left[-\frac{\sigma_T(z\cos\theta)^2+\sigma_M^2(z\sin\theta)^2}{2\sigma_M^2\sigma_T^2}\right]d\theta$$

$$= \frac{z}{2\pi\sigma_M\sigma_T}\int_0^{2\pi}\exp\left[-z^2\frac{\left(\sigma_T^2-\sigma_M^2\right)^2\cos^2\theta+\sigma_M^2}{2\sigma_M^2\sigma_T^2}\right]d\theta$$

$$= \frac{z}{2\pi\sigma_M\sigma_T}\exp\left(-\frac{z^2\sigma_M^2}{2\sigma_M^2\sigma_T^2}\right)\int_0^{2\pi}\exp\left[-\frac{z^2\left(\sigma_T^2-\sigma_M^2\right)}{2\sigma_M^2\sigma_T^2}\cos^2\theta\right]d\theta$$

With

$$\cos^2\theta = \frac{1+\cos^2\theta}{2},$$

we get

$$f_Z(z) = \frac{z}{2\pi\sigma_M\sigma_T}\exp\left(-\frac{z^2\sigma_M^2}{2\sigma_M^2\sigma_T^2}\right)\int_0^{2\pi}\exp\left[-\frac{z^2\left(\sigma_T^2-\sigma_M^2\right)^2}{4\sigma_M^2\sigma_T^2}(1+\cos 2\theta)\right]d\theta$$

Denoting

$$2\theta = \psi, \quad 2d\theta = d\psi$$

we have

$$f_Z(z) = \frac{z}{4\pi\sigma_M\sigma_T}\exp\left(-\frac{z^2\sigma_M^2}{2\sigma_M^2\sigma_T^2}\right)\int_0^{4\pi}\exp\left[-\frac{z^2\left(\sigma_T^2-\sigma_M^2\right)}{4\sigma_M^2\sigma_T^2}(1+\cos\psi)\right]d\psi$$

$$= \frac{z}{4\pi\sigma_M\sigma_T}\exp\left[-\frac{2z^2\sigma_M^2+z^2\left(\sigma_T^2-\sigma_M^2\right)}{4\sigma_M^2\sigma_T^2}\right]$$

$$\times\int_0^{4\pi}\exp\left[-\frac{z^2\left(\sigma_T^2-\sigma_M^2\right)}{4\sigma_M^2\sigma_T^2}\cos\psi\right]d\psi$$

Integral equals

$$\int_0^{2\pi}\exp\left[-\frac{z^2\left(\sigma_T^2-\sigma_M^2\right)}{4\sigma_M^2\sigma_T^2}\cos\psi\right]d\psi+\int_{2\pi}^{4\pi}\exp\left[-\frac{z^2\left(\sigma_T^2-\sigma_M^2\right)}{4\sigma_M^2\sigma_T^2}\cos\psi\right]d\psi$$

In the second integral we substitute $\psi - 2\pi = \kappa$, and

$$\int_{2\pi}^{4\pi} \exp\left[-\frac{z^2\left(\sigma_T^2 - \sigma_M^2\right)}{4\sigma_M^2\sigma_T^2} cos\,\psi\right] d\psi$$
$$= \int_{0}^{2\pi} \exp\left[-\frac{z^2\left(\sigma_T^2 - \sigma_M^2\right)}{4\sigma_M^2\sigma_T^2} \cos\left(\kappa + 2\pi\right)\right] d\kappa$$

but $\cos(\kappa + 2\pi) = \cos\kappa$, and therefore the original integral becomes

$$\int_{0}^{4\pi} \exp\left[-\frac{z^2\left(\sigma_T^2 - \sigma_M^2\right)}{4\sigma_M^2\sigma_T^2} \cos\psi\right] d\psi = 2\int_{0}^{2\pi} \exp\left[-\frac{z^2\left(\sigma_T^2 - \sigma_M^2\right)}{4\sigma_M^2\sigma_T^2} cos\,\psi\right] d\psi$$

But, in view of Equation 7.46

$$I_0\left(x\right) = \frac{1}{2\pi}\int_{0}^{2\pi} e^{x\cos\theta}\,d\theta$$

Hence

$$2\int_{0}^{2\pi} \exp\left[-\frac{z^2\left(\sigma_T^2 - \sigma_M^2\right)}{4\sigma_M^2\sigma_T^2} \cos\psi\right] d\psi$$
$$= 4\pi\, I_0\left[\frac{z^2\left|\sigma_M^2 - \sigma_T^2\right|}{4\sigma_M^2\sigma_T^2}\right]$$

and

$$fz\left(z\right) = \frac{z}{\sigma_M\sigma_T} \exp\left[-\frac{z^2\left(\sigma_M^2\sigma_T^2\right)}{2\sigma_M^2\sigma_T^2}\right] I_0\left[\frac{z^2\left|\sigma_M^2 - \sigma_T^2\right|}{4\sigma_M^2\sigma_T^2}\right], \quad \text{Q.E.D.}$$

Now we determine the reliability: Let us first find $F_Z(z)$.

$$F_Z\left(z\right) = \frac{1}{\sigma_M\sigma_T}\int_{0}^{z} z e^{-\frac{z^2}{a^2}} I_0\left(\frac{z}{b^2}\right) dz$$

Whence

$$a^2 = \frac{2\sigma_M^2\sigma_T^2}{\sigma_M^2 + \sigma_T^2}, \quad b^2 = \frac{4\sigma_M^2\sigma_T^2}{\left|\sigma_M^2 - \sigma_T^2\right|}$$

Consider the case when $\sigma_M = \sigma_T = \sigma$. Then

$$fz\left(z\right) = \frac{z}{\sigma^2} \exp\left(-\frac{z^2}{\sigma^2}\right)$$

or becomes Rayleigh distributed random variable. Hence

$$F_Z(z) = 1 - \exp\left(-\frac{z^2}{2\sigma^2}\right)$$

Since the strength requirement is

$$\sum_{eq} = \frac{M_{eq}}{S} \leq \sigma_{\text{YIELD}}$$

we have $z = \sigma_{\text{YIELD}}\, S$ and

$$R = 1 - \exp\left(-\frac{\sigma_{YIELD}^2 S^2}{2\sigma^2}\right)$$

PROBLEM 7.5

Extend Example 7.7 to the case where the von Mises stress theory of failure is used instead of the maximum shear stress theory.

SOLUTION 7.5

In case of the maximum shear stress theory of failure the equivalent moment is

$$Z = M_{eq} = (M^2 + T^2)^{1/2}$$

For the case of the von Mises criterion (maximum distortion energy criterion) we have:

$$Z = M_{eq} = \left(M^2 + \frac{3}{4}T_2\right)^{1/2}$$

we introduce the auxiliary variable

$$P = \frac{\sqrt{3}}{2}T$$

so that

$$f_{MP}(m, p) = f_{MT}\left(m, \frac{2}{\sqrt{3}}p\right)\left|\frac{\partial(m, t)}{\partial(m, p)}\right|$$

with

$$\frac{\partial(m,t)}{\partial(m,p)} = \begin{vmatrix} 1 & 0 \\ 0 & \frac{2}{\sqrt{3}} \end{vmatrix} = \frac{2}{\sqrt{3}}$$

and

$$f_{MP}(m, p) = \frac{2}{\sqrt{3}} f_{MT}\left(m, \frac{2}{\sqrt{3}}P\right)$$

Now we introduce, following Example 7.7, auxiliary variables Z and θ, so that $M = Z\cos\theta,\ P = Z\sin\theta$

with

$$f_{Z\theta}(z,\theta) = z\, f_{MP}(z\,\cos\theta, z\sin\theta) = z\frac{2}{\sqrt{3}} f_{MT}\left(z\cos\theta, \frac{2}{\sqrt{3}} z\sin\theta\right)$$

Now

$$f_Z(z) = \frac{2}{\sqrt{3}} z \int_0^{2\pi} f_{MT}\left(z\cos\theta, \frac{2}{\sqrt{3}} z\sin\theta\right) d\theta \tag{7.1}$$

and

$$F_Z(z) = \int_0^z f_Z(z)\, dz$$

Calculations for M which is $N(a, \sigma_1^2)$ and T which is $N(b, \sigma_1^2)$ and independent are performed in the similar manner as in Example 7.7.

PROBLEM 7.6

The clamped-clamped beam is subjected to a load as shown in the accompany figure. Both Q and $\sum_{\text{allow}}$ have log-normal distributions. Find the reliability of the beam.

SOLUTION 7.6

The maximum bending moment reads

$$M_{max} = \frac{Ql^2}{12}$$

(see e.g., E. Popov, "Introduction to Mechanics of Solids").

The strength requirement is

$$\sum = \frac{M_{max}}{S} \le \sum_Y,\ \sum_Y = \sum_{\text{YIELD}}$$

since both $\sum$ and $\sum_Y$ assume positive values. Now

$$S = \frac{\pi c^3}{4}$$

and

$$\frac{Ql^2}{12\cdot(\pi c^3/4)} \le \sum_Y,\ \frac{Q}{\sum_Y} \le \frac{3\pi c^2}{l^2}$$

Assume that Q and $\sum_Y$ are independent. Then

$$Z \equiv lnQ - ln\sum_Y \leq \ln(3\pi c^3/l^2)$$

and

$$E(Z) = E(lnQ) - E\left(ln\sum_Y\right) = e^{E(Q)+\frac{1}{2}\sigma_Q^2 - e^{-E(\sum_Y)+\frac{1}{2}\sigma\sum_Y^2}}$$

$$Var(Z) = Var\,(lnQ) + Var\left(\sum_Y\right)$$

$$= \left\{\exp\left[2E\,(Q) + \sigma_Q^2\right]\right\}\left[\exp\left(\sigma_Q^2\right) - 1\right]$$

$$+ \left\{\exp\left[2E\left(\sum_Y\right) + \sigma_{\sum_Y^2}\right]\right\}\left[\exp\left(\sigma_{\sum_Y^2}\right) - 1\right]$$

Since Z is normally distributed we get

$$R = P[Z \leq \ln(3\pi c^3/l^2)]$$

$$= \frac{1}{2} + \text{erf}\left[\frac{\ln(3\pi c^3/l^2) - E(Z)}{\sqrt{Var(Z)}}\right]$$

where $E(Z)$ and $Var(Z)$ are as above.

PROBLEM 7.7

The cantilever is subjected to a pair of random independent moments with gamma distributions

$$f_{M_1}(m_1) = \frac{1}{\lambda^{\alpha+1}\Gamma(\alpha+1)} m_1^{\alpha} e^{-m_1/\lambda} U(m_1)$$

$$f_{M_2}(m_2) = \frac{1}{\lambda^{\beta+1}\Gamma(\beta+1)} m_2^{\beta} e^{-m_2/\lambda} U(m_2)$$

Show that the maximum moment also has a gamma distribution, and calculate the reliability.

SOLUTION 7.7

The maximum bending moment appears at the clamped end on the cantilever is

$$M = M_1 + M_2$$

If $M_{M_1}(\theta_1)$ is a characteristic function of M_1, and $M_{M_2}(\theta_2)$ that of M_2, then, according with Equation 6.99

$$M_{M_1+M_2}(\theta) = E\{\exp[\theta_1 M_1 + \theta_2 M_2]\};$$

For independent M_1 & M_2:

$$M_{M_1+M_2}(\theta) = M_{M_1}(\theta) M_{M_2}(\theta)$$

Now

$$M_{M_1}(\theta) = \frac{1}{(1 - i\theta_1\lambda)^{\alpha+1}}, \quad M_{M_2}(\theta_2) = \frac{1}{(1 - i\theta_2\lambda)^{\beta+1}}$$

Therefore

$$M_{M_1+M_2}(\theta) = \frac{1}{(1 - i\theta_1\lambda)^{\alpha+1}} \frac{1}{(1 - i\theta_2\lambda)^{\beta+1}} = \frac{1}{(1 - i\theta\lambda)^{\alpha+\beta+2}}$$

Due to the unique correspondence between the probability density and the characteristic function, $M_1 + M_2$ is also gamma-distributed with parameters λ and $\alpha + \beta + 1$

$$f_M(m) = \begin{cases} 0, & m < 0 \\ \lambda^{\alpha+\beta+2}\Gamma^{(\alpha+\beta+2)} m^{\alpha+\beta+1} e^{-m/\lambda}, & m > 0 \end{cases}$$

(check this solution with the convolution integral calculation).

If $\alpha + \beta$ is integer then

$$f_M(m) = \frac{1}{\lambda^{\alpha+\beta+2}(\alpha + \beta + 1)!} m^{\alpha+\beta+1} e^{-\frac{m}{\lambda}} U(m)$$

where $U(\ldots)$ is a unit step function. Then

$$F_M(m) = 1 - e^{-\frac{m}{\lambda}} \sum_{k=0}^{\alpha+\beta+1} \frac{1}{k!} \left(\frac{x}{\lambda}\right)^k$$

Since strength requirement is

$$\sum = \frac{M_{max}}{S} \leq \sigma_Y$$

where σ_Y is a yield stress and S the section modulus, we get

$$M_{max} \leq S\sigma_Y$$

or

$$R = F_M(\sigma_Y S) = 1 - e^{-\sigma_Y S/\lambda} \sum_{k=0}^{\alpha+\beta+1} \frac{1}{k!}\left(\frac{\sigma_Y S}{\lambda}\right) k$$

If, say,

$$\alpha = 2, \quad \beta = 1, \quad \lambda = 10, \quad S\sigma_Y = 60$$

then

$$R = 1 - \sum_{k=0}^{4} e^{-\frac{60}{10}} \left(\frac{60}{10}\right)^k \frac{1}{k!}$$

$$= 1 - e^{-6}\left(1 + 6 + \frac{6^2}{2!} + \frac{6^3}{3!} + \frac{6^4}{4!}\right) = 0.7149.$$

Let us check what is obtained by the central limit theorem approximation (see pp. 262–264). Under it, the sum $M_1 + M_2$ is approximately normally distributed with

$$E(M) = E(M_1) + E(M_2) = (\alpha + 1)\lambda + (\beta + 1)\lambda = (\alpha + \beta + 2)\lambda$$

$$Var(M) = Var(M_1) + Var(M_2) = (\alpha + 1)\lambda^2 + (\beta + 1)\lambda^2 = (\alpha + \beta + 2)\lambda^2$$

in virtue of equations appearing in the first line of pp. 77. Then

$$R \simeq \frac{1}{2} + \text{erf}\left[\frac{S\sigma_Y - (\alpha + \beta + 2)\lambda}{\sqrt{\alpha + \beta + 2\lambda}}\right]$$

or, with actual data

$$R \simeq \frac{1}{2} + \text{erf}\left[\frac{60 - (2 + 2 + 2) \cdot 10}{\sqrt{2 + 2 + 2 \cdot 10}}\right]$$

$$= \frac{1}{2} + \text{erf}(0) = \frac{1}{2}$$

i.e. since only two constituent moments are involved, the approximation, based on the central limit theorem, is unsatisfactory.

PROBLEM 7.8

For the system of Problem 7.7, assume that the moments have identical Rayleigh distribution

$$f_{M_1}(x) = f_{M_2}(x) = \frac{x}{a^2}\exp(-x^2/2a^2)U(x)$$

Verify that the reliability of the cantilever is

$$R = 1 - \exp(-z^2/2a^2) - \sqrt{\pi}\frac{z}{a}\exp(-z^2/4a^2)\mathrm{erf}\left(\frac{z}{a\sqrt{2}}\right)$$

where $z = \sigma_{\text{allow}}S$.

SOLUTION 7.8

We use Equation (6.51),

$$F_M(y) = \int_{-\infty}^{\infty} F_{X_1}(y - x_2)\, f_{X_2}(x_2)\, dx_2$$

Now

$$F_{X_1}(x_1) = [1 - \exp(-x_1^2/2a^2)]U(x)$$

Hence

$$F_{X_1}(y - x_2) = \{1 - \exp[-(y - x_2)^2/2a^2]\}\frac{x}{a^2}\exp\left(-\frac{x^2}{2a^2}\right)dx$$

$$= \int_0^y \frac{x}{d^2}\exp\left(-\frac{x^2}{2a^2}\right)dx - \int_0^y \frac{x}{a^2}\exp\left(-\frac{y^2}{4a^2}\right)$$

$$\times \exp\left[-\frac{(y/\sqrt{2} - \sqrt{2}x)^2}{2a^2}\right]dx$$

$$= 1 - \exp\left(-\frac{y^2}{2a^2}\right) - \frac{1}{a^2}\exp\left(-\frac{y^2}{4a^2}\right)$$

$$\times \int_0^y x exp\left[-\frac{(y/\sqrt{2} - \sqrt{2}x)^2}{2a^2}\right]dx$$

Let

$$I = \int_0^y x exp\left[-\frac{(y/\sqrt{2} - \sqrt{2}x)^2}{2a^2}\right]dx$$

We denote

$$y/\sqrt{2} - \sqrt{2}x = u$$

$$x = (y/\sqrt{2} - u)\frac{1}{\sqrt{2}}, \quad dx = -\frac{du}{\sqrt{2}}$$

$$x = 0 : u = y/\sqrt{2}$$

$$x = y : u = -y/\sqrt{2}$$

As a result

$$\begin{aligned} I &= \int_{y/\sqrt{2}}^{-y/\sqrt{2}} \frac{1}{\sqrt{2}}\left(\frac{y}{\sqrt{2}} - u\right)\exp\left(-\frac{u^2}{2a^2}\right)\frac{du}{(-\sqrt{2})} \\ &= \int_{y/\sqrt{2}}^{-y/\sqrt{2}} \frac{y}{\sqrt{2}}\exp\left(-\frac{u^2}{2a^2}\right)du + \int_{y/\sqrt{2}}^{-y/\sqrt{2}} \frac{u}{2}\exp\left(-\frac{u^2}{2a^2}\right)du \\ &\equiv I_1 + I_2 \end{aligned}$$

We calculate now I_1. Let $v = u/a; du = adu$.

$$\begin{aligned} I_1 &= \int_{y/\sqrt{2}}^{-y/\sqrt{2}} \frac{ay}{\sqrt{2}}\exp\left(-\frac{v^2}{2}\right)dv \\ &= -\frac{ay}{\sqrt{2}}\sqrt{2\pi}\left[\operatorname{erf}\left(-\frac{y}{a\sqrt{2}}\right) - \operatorname{erf}\left(\frac{y}{a\sqrt{2}}\right)\right] \\ &= \frac{ay\sqrt{\pi}}{2}\cdot 2\operatorname{erf}\left(\frac{y}{a\sqrt{2}}\right) = ay\sqrt{\pi}\operatorname{erf}\left(\frac{y}{a\sqrt{2}}\right) \end{aligned}$$

For calculation of the integral I_2 we note that the integrand is an odd function of u, whereas the bounds are symmetric; therefore, I_2 vanishes identically. Final result for I is

$$I = ay\sqrt{\pi}\operatorname{erf}(y/a\sqrt{2})$$

This results in $F_Y(y)$

$$F_Y(y) = 1 - \exp\left(-\frac{y^2}{2a^2}\right) - \exp\left(-\frac{y^2}{4a^2}\right)\frac{y\sqrt{\pi}}{a}\operatorname{erf}\left(\frac{y}{a\sqrt{2}}\right), \quad \text{Q.E.D.}$$

PROBLEM 7.9

A cantilever is subjected to three concentrated moments having uniform distribution in the interval (0, 1). Show that

$$f_M(m) = \begin{cases} 0, & m < 0 \\ \frac{1}{2}m^2, & 0 \le m < 1 \\ \frac{1}{2}m^2 - \frac{3}{2}(m-1)^2, & 1 \le m < 2 \\ \frac{1}{2}m^2 - \frac{3}{2}(m-1)^2 + \frac{3}{2}(m-2)^2, & 2 \le m < 3 \\ 0, & m \ge 3 \end{cases}$$

so that the graph consists of segments of three different parabolas, in the interval (0, 3). Find the reliability of the structure, and compare with the estimate by the central limit theorem.

SOLUTION 7.9

A cantilever is subjected to three concentrated moments having uniform distribution in the interval (0, 1). Show that

Consider first the two moments

$$Y = M_1 + M_2$$

with

$$f(m_1, m_2) = \begin{cases} 1, & \text{for } 0 < m_1 < 1 \text{ and } 0 < m_2 < 1 \\ 0, & \text{otherwise} \end{cases}$$

We utilize formula 6.52:

$$f_Y(y) = \int_{-\infty}^{\infty} f_{M_1}(y - M_2)\, f_{M_2}(m_2)\, dM_2$$

Now

$$f(y - m_2)\, f_{M_2}(m_2) = \begin{cases} 1, & \text{for } 0 < m_2 < 1,\ y = m_1 + m_2 > m_2 \text{ and} \\ & m_1 = y - m_2 < 1 \text{ (shaded in Figure)} \\ & 0, \text{ elsewhere} \end{cases}$$

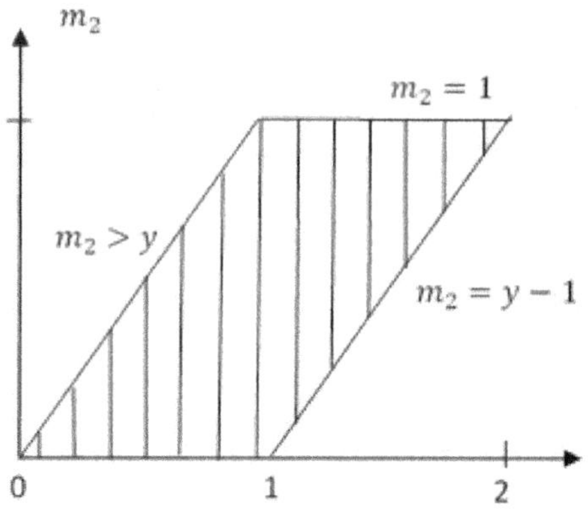

Finally

$$f_M(m) = \frac{1}{2} - \frac{(m-1)^2}{2} - \frac{(2-m)^2}{2} + \frac{(2-1)^2}{2}$$

$$= 1 - \frac{(m-1)^2}{2} - \frac{4-4m+m^2}{2}$$

$$= \frac{2-m^2+2m-1-4+4m+m^2}{2} = \frac{-3-2m^2+6m}{2}$$

$$= \frac{m^2-(3m^2-6m+3m)}{2}\frac{m^2-3(m^2-2m+1)}{2}$$

$$= \frac{m^2}{2} - \frac{3}{2}(m-1)^2$$

For $2 \le m \le 3$

$$f_M(m) = \int_{m-1}^{2} (2-y)\,dy$$

$$= 2y - \frac{1}{2}y^2\Big]_{m-1}^{2} = 4 - 2(m-1) - \frac{2^2}{2} + \frac{(m-1)^2}{2}$$

$$= 4 - 2m + \frac{(m-1)^2}{2} = \frac{m^2-6m+9}{2}$$

$$= \frac{m^2}{2} - \frac{3}{2}(m-1)^2 + \frac{3}{2}(m-2)^2$$

for $m > 3$

$$f_M(m) = \int_2^{\infty} 0\,dy = 0$$

PROBLEM 7.10

Prove Eq. (7.61).

Note to lecturer: the problem should read: Prove Equation (7.60).

SOLUTION 7.10

By independence

$$f_{M_1M_2}(m_1, m_2) = f_{M_1}(m_1) f_{M_2}(m_2)$$

and

$$F_M(m) = \text{Prob}(M_1 + M_2 \le m) = \iint_{m_1+m_2\le m} (m_1)\, f_{M_2}(m_2)\, dm_1 dm_2$$

If $0 < m < 1$ we have

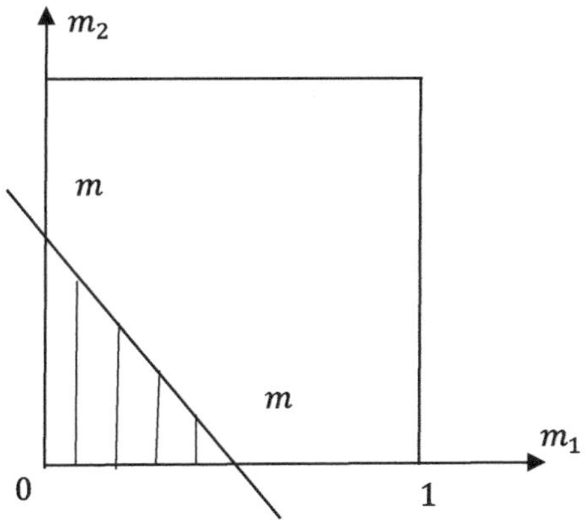

$$F_M(m) = \iint 1 \cdot dm_1 dm_2 = \text{shaded area} = m^2/2$$

If $1 \le m \le 2$, we have

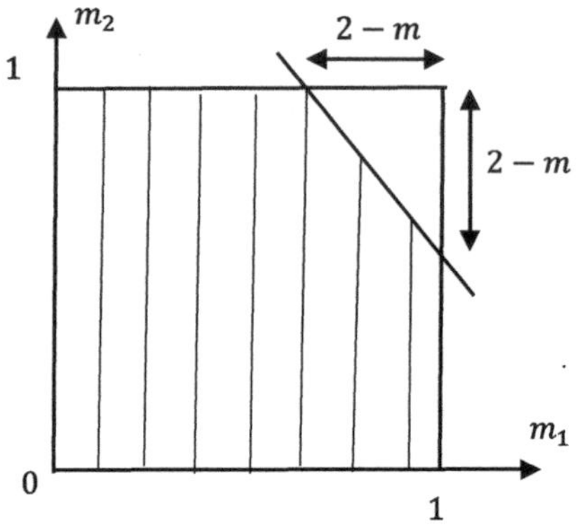

$$F_M\,(m) = \text{shaded area} = 1 - \frac{(2-m)^2}{2}$$

Thus also, $f_M(m) = m,\ 0 \le m \le 1;\ f_M(m) = 2 - m,\ 1 \le m \le 2,\ f_M(m) = 0$ elsewhere.

PROBLEM 7.11

The probability density (7.60) is referred to as a *generalized Erlang density of the first order.* Show by mathematical induction that the sum of n independent random variables, each of which has an exponential distribution with different parameters $\lambda_1,\ \lambda_2, \ldots, \lambda_n$, has a generalized Erlang distribution of $(n-1)$th order, given by

$$f_{n-1}\,(m) = (-1)^{n-1} \prod_{i=1}^{n} \lambda_1 \sum_{j=1}^{n} \frac{\exp(-\lambda_j m)}{\prod_{k\neq j}^{n} (\lambda_j - \lambda_k)}$$

where the product $\prod_{k\neq j}^{n}$ is taken over $k = 1, 2, \ldots, j-1, j+1, \ldots, n$. The distribution function of M is

$$F_M(m) = (-1)^{n-1} \prod_{i=1}^{n} \lambda_i \sum_{j=1}^{n} \frac{1-\exp(-\lambda_j m)}{\lambda_j \prod_{k\neq j} (\lambda_j - \lambda_k)}.$$

Note to lecturer: The first sentence should read: "The probability density (7.66) is referred to as a generalized Erlang density of the first order."

SOLUTION 7.11

The formula given in the problem yields for $n = 2$:

$$f_1(m) = \lambda_1\lambda_2\left(\frac{e^{-\lambda_1 m}}{\lambda_1-\lambda_2}+\frac{e^{-\lambda_2 m}}{\lambda_2-\lambda_1}\right)$$

$$= \lambda_1\lambda_2\left(\frac{e^{-\lambda_1 m}-e^{-\lambda_2 m}}{\lambda_1-\lambda_2}\right)$$

as given in Equation (7.65).

The appropriate distribution function reads, for the generalized Erlang distribution of $(n-1)$th order:

$$F_{n-1}(m) = (-1)^{n-1}\prod_{i=1}^{n}\lambda_i\sum_{j=1}^{n}\frac{1-e^{-\lambda_j m}}{\sum_{j\neq k}^{n}(\lambda_j-\lambda_k)}$$

In particular case $\lambda_1 = \lambda_2 = \cdots \lambda_n = \lambda$ we get Erlang's distribution of $(n-1)$th order

$$f_{n-1}(m) = \frac{\lambda(\lambda m)^{n-1}}{(n-1)!}$$

and

$$F_M(m) = \int_0^m f_{n-1}(m)\,dm = 1 - \int_m^\infty f_{n-1}(m)\,dm = 1 - R(n-1,\lambda m)$$

where

$$P(m,n) = \frac{a^m}{m!}e^{-a},\quad R(m,a) = \sum_{k=0}^{m}\frac{a^k}{k!}e^{-a}$$

The case of three random variables is tackled as considering the sum

$$f_{M_3}(m_3) = \lambda_3 e^{-\lambda_3 m_3}$$

$$f_Z(z) = \lambda_1\lambda_2\frac{e^{-\lambda_1 z}-e^{-\lambda_2 z}}{\lambda_1-\lambda_2}$$

with

$$M = Z + M_3$$

Then

$$
\begin{aligned}
f_M(m) &= \int_{-\infty}^{\infty} f_Z(x)\, f_{M_3}(m-x)\, dx \\
&\times \int_0^m \frac{\lambda_1\lambda_2}{\lambda_1-\lambda_2}\left(e^{-\lambda_1 x}-e^{-\lambda_2 x}\right)\lambda_3 e^{-\lambda_3(m-x)} dx \\
&= \frac{\lambda_1\lambda_2\lambda_3}{\lambda_1-\lambda_2} e^{-\lambda_3 m} \int_0^m \left[e^{(\lambda_3-\lambda_1)x}-e^{(\lambda_3-\lambda_2)x}\right] dx \\
&= \frac{\lambda_1\lambda_2\lambda_3}{\lambda_1-\lambda_2} e^{-\lambda_3 m}\left[\frac{e^{(\lambda_3-\lambda_1)m}-1}{\lambda_3-\lambda_1}-\frac{e^{-(\lambda_3-\lambda_2)m}-1}{\lambda_3-\lambda_2}\right] \\
&= \frac{\lambda_1\lambda_2\lambda_3}{\lambda_1-\lambda_2}\left[\frac{e^{-\lambda_1 m}-e^{-\lambda_3 m}}{\lambda_3-\lambda_1}-\frac{e^{-\lambda_2 m}-e^{-\lambda_3 m}}{\lambda_3-\lambda_2}\right] \\
&= (-1)^3\lambda_1\lambda_2\lambda_3 \\
&\times \left[\frac{e^{-\lambda_1 m}}{(\lambda_1-\lambda_2)(\lambda_1-\lambda_3)}+\frac{e^{-\lambda_2 m}}{(\lambda_2-\lambda_1)}+\frac{e^{-\lambda_3 m}}{(\lambda_3-\lambda_1)(\lambda_3-\lambda_2)}\right]
\end{aligned}
$$

which is in accordance with the formula which should be proved. It is easy then to show that if it is valid for $n = k$, it is so for $n = k + 1$, by using convolution integral.

PROBLEM 7.12

Assume M and T in Example 7.7 to be uniformly distributed in the circle $m^2+t^2 \leq p^2$ (see also Prob. 6.1), that is,

$$
f_{MT}(m,t) = \begin{cases} \dfrac{1}{\pi p^2}, & m^2+t^2 \leq p^2 \\ 0, & \text{otherwise} \end{cases}
$$

Find the probability distribution and density functions of the maximum shear stress $\tau_{\max}$. Find the reliability of the circular shaft.

SOLUTION 7.12

In accordance with the formula preceding (7.43),

$$f_Z(z) = zU(z)\int_0^{2\pi} f_{MF}(z\cos\theta, z\sin\theta)\,d\theta, \quad z \le p$$

we get since $f_{MT} = \text{const}$,

$$f_Z(z) = \begin{cases} z\,U(z)\,\dfrac{2\pi}{\pi p^2}, & z \le p \\ 0, & \text{otherwise} \end{cases}$$

Therefore

$$f_Z(z) = \int_0^z f_Z(z)\,dz = \begin{cases} \dfrac{z^2}{p^2}, & \text{for } z \le p \\ 1, & \text{for } z > p \end{cases}$$

and

$$R = F_Z(\sigma_Y S) = F_Z\left(\sigma_Y \frac{\pi}{4} c^3\right) = \begin{cases} \dfrac{\pi^2\sigma_Y^2 c^6}{16}, & \text{for } \dfrac{\pi}{4}\sigma_Y c^3 \le p \\ 1, & \text{for } \dfrac{\pi}{4}\sigma_Y c^3 > p \end{cases}$$

If required reliability is r, we solve the design problem as follows:

$$\frac{\pi^2\sigma_Y^2}{16} c_{\text{req}}^6 = r$$

and get

$$C_{\text{req}} = \sqrt[6]{\frac{16r}{\pi^2\sigma_Y^2}}$$

chapter 8

Elements of the Theory of Random Functions

PROBLEM 8.1

Given a pair of random functions

$$X(t) = A \sin \omega_1 t, \quad Y(t) = B \sin \omega_2 t$$

where A and B are random variables with mathematical expectations $E(A) = 1$, $E(B) = 2$, and the variance-covariance matrix

$$\begin{bmatrix} 1 & 1 \\ 1 & 9 \end{bmatrix}$$

Determine the autocorrelation functions $R_X(t_1, t_2)$, $R_Y(t_1, t_2)$ and the cross-correlation functions $R_{XY}(t_1, t_2)$, $R_{YX}(t_1, t_2)$.

SOLUTION 8.1

The autocorrelation functions are

$$R_X(t_1, t_2) = E\{[A \sin \omega_1 t_1 - E(A \sin \omega_1 t_1)][A_1 \sin \omega_1 t_2 - E(A \sin \omega_1 t_2)]\}$$

but

$$E(A \sin \omega_1 t) = E(A) \sin \omega_1 t$$

and

$$\begin{aligned}R_X(t_1, t_2) &= E\{A - E(A)][A - E(A)]\} \sin \omega_1 t_1 \sin \omega_1 t_2 \\ &= Var(A) \sin \omega_1 t_1 \sin \omega_1 t_2 = 1 \cdot \sin \omega_1 t_1 \sin \omega_1 t_2 \\ &= \sin \omega_1 t_1 \sin \omega_1 t_2\end{aligned}$$

Analogously

$$R_Y(t_1, t_2) = Var(B) \sin \omega_2 t_1 \sin \omega_2 t_2 = 9 \sin \omega_2 t_1 \sin \omega_2 t_2$$

Now

$$\begin{aligned}R_{XY}(t_1, t_2) &= E\{[A - E(A)] \sin \omega_1 t_1 [B - E(B) \sin \omega_2 t_2]\} \\ &= E\{[A - E(A)][B - E(B)]\} \sin \omega_1 t_1 \sin \omega_2 t_2 \\ &= \text{Cov}(A, B) \sin \omega_1 t_1 \sin \omega_2 t_2 \\ &= 1 \cdot \sin \omega_1 t_1 \sin \omega_2 t_2 \\ R_{YX}(t_1, t_2) &= \sin \omega_2 t_1 \sin \omega_1 t_2\end{aligned}$$

PROBLEM 8.2

Solve Problem 8.1 with the variance-covariance matrix

$$\begin{bmatrix} 1 & 3-\varepsilon \\ 3-\varepsilon & 9 \end{bmatrix}$$

for $\varepsilon \leq 1$. What happens when $\varepsilon \to 0$?

SOLUTION 8.2

Under new circumstances $R_X(t_1, t_2)$ and $R_Y(t_1, t_2)$ remain unchanged. The cross-correlation functions become

$$\begin{aligned}R_{XY}(t_1, t_2) &= E\{[A - E(A)]B - E(B)]\} \sin \omega_1 t_1 \sin \omega_2 t_2 \\ &= (3 - \varepsilon) \sin \omega_1 t_1 \sin \omega_2 t_2 \\ R_{YX}(t_1, t_2) &= (3 - \varepsilon) \sin \omega_2 t_1 \sin \omega_1 t_2\end{aligned}$$

at $t_1 = t_2 = t$ we get

$$\begin{aligned}R_X(t, t) &= \sin \omega_1 t \sin \omega_1 t = (\sin \omega_1 t^2) \\ R_Y(t, t) &= 9(\sin \omega_2 t)^9\end{aligned}$$

$$R_{XY}(t,t) = (3-\varepsilon)\sin\omega_1 t \sin\omega_2 t$$

$$R_{YX}(t,t) = (3-\varepsilon)\sin\omega_1 t \sin\omega_2 t$$

In a particular case $\omega_1 = \omega_2$ we get

$$[R_1(t,t)] = \begin{bmatrix} 1 & 3-\varepsilon \\ 3-\varepsilon & 9 \end{bmatrix} (\sin\omega_1 t)^2$$

The determinant of the appropriate matrix is

$$9-(9-\varepsilon)^2$$

and vanishes for $\varepsilon \to 0$. This implies that the random variables (for fixed t) $X(t)$ and $Y(t)$ become dependent (see pp. 204–205).

PROBLEM 8.3

Check whether a random function $X(t)$ possessing the autocorrelation function

$$R_X(\tau) = d^2 e^{-\alpha|\tau|}(1+\alpha|\tau|)$$

if differentiable or otherwise. Compare with the result obtained for $R_X(\tau)$ given in Eq. (8.106), and offer an interpretation.

SOLUTION 8.3

The spectral density $S_x(\omega)$ of the random function is

$$S_X(\omega) = \omega^2 S_X(\omega) = \frac{2\alpha^3 d^2}{\pi}\frac{\omega^2}{(\alpha^2+\omega^2)^2} \tag{1}$$

The spectral density of X' is

$$S_X(\omega) = \omega^2 S_X(\omega) = \frac{2\alpha^3 d^2}{\pi}\frac{\omega^2}{(\alpha^2+\omega^2)^2} \tag{2}$$

Due to the fact that $S_X(\omega)$ decays faster than ω^{-3}, the integral

$$\int_{-\infty}^{\infty} S_X(\omega)d\omega < \infty$$

and therefore $X(t)$ is a differentiable random function. Indeed, this result can be compared with that for the random function with the autocorrelation function as per (8.106):

$$R_X(\tau) = d^2 e^{-\alpha|\tau|}(\cos\beta\tau + \gamma \sin\beta|\tau|) \tag{3}$$

for $\gamma = \frac{\alpha}{\beta}$. For small values of β we get

$$\begin{aligned}
\cos\beta\tau &\simeq 1, \ \sin\beta|\tau| \approx \beta|\tau| \ \text{ and} \\
R_X(\tau) &= d^2 e^{-\alpha|\tau|}\left(\cos\beta\tau + \frac{\alpha}{\beta}\sin\beta|\tau|\right) \\
&\rightarrow d^2 e^{-\alpha|\tau|}(1 + \alpha|\tau|)
\end{aligned} \tag{4}$$

Since process with autocorrelation function (3) is differentiable (for $\gamma \leq \alpha/\beta$) so should be that with autocorrelation function (4). The latter being obtainable from Eq. (3).

PROBLEM 8.4

Show that for a weakly stationary random function $X(t)$,

$$S_{X^{(n)}}(\omega) = (-1)^n (i\omega)^{2n} S_X(\omega)$$

SOLUTION 8.4

Indeed, according to Equation 8.104,

$$S_{X'}(\omega) = \omega^2 S_X(\omega)$$

Now, in complete analogy

$$S_{X^{(n)}}(\omega) = \omega^2 S_X(n-1)^{(\omega)}$$

which leads to the desired result.

PROBLEM 8.5

Verify that the X' with $R_X(\tau)$ as in Problem 8.3 is

$$= d^2 e^{-\alpha|\tau|}(1 - \alpha|\tau|)$$

Note to lecturer: The problem should read: Verify that the autocorrelation function $R_X(\tau)$ of the X' with $R_X(\tau)$ as in Problem 8.3 is $R_{X'}(\tau) = d^2\exp(-\alpha|\tau|)(1-\alpha|\tau|)$

SOLUTION 8.5

$$R_Y(\tau) = -\frac{d^2 R_x(\tau)}{d\tau^2} = -\frac{d^2}{d\tau^2}[d^2 e^{-\alpha|\tau|}(1+\alpha|\tau|)]$$

But

$$\frac{d|\tau|}{d\tau} = \text{sign}\tau$$

Therefore

$$R_Y(\tau) = d^2\alpha^2 \frac{d}{d\tau}[e^{-\alpha|\tau|}\text{sign}\tau \cdot |\tau|]$$

On the other hand

$$\text{sign}\tau \cdot |\tau| = \tau$$

and

$$R_Y(\tau) = d^2\alpha^2 \frac{d}{d\tau}[e^{-\alpha|\tau|}\tau]$$

$$= d^2\alpha^2[e^{-\alpha|\tau|} - \alpha e^{-\alpha|\tau|}\tau \frac{d|\tau|}{d\tau} = d^2\alpha^2 e^{-\alpha|\tau|}(1-\alpha|\tau|)$$

PROBLEM 8.6

Check whether $X(t)$ in Problem 8.5 has a second derivative.

SOLUTION 8.6

The spectral density of $X(t)$ is

$$S_X(\omega) = \frac{2\alpha^3 d^2}{\pi(\alpha^2+\omega^2)^2}$$

The second derivative's spectral density would be

$$S_{X''}(\omega) = \omega^4 S_X(\omega)$$

$$S_{X''}(\omega) = \frac{2\alpha^3 d^2}{\pi} \frac{\omega^4}{(\alpha^2+\omega^2)^2}$$

But

$$\int_{-\infty}^{\infty} \frac{\omega^4}{(\alpha^2+\omega^2)^2} \to \infty$$

Therefore $X(t)$ has no second derivative.

On the other hand the process $X(t)$ with autocorrelation function

$$R_X(\tau) = d^2 e^{-\alpha|\tau|}\left(1+\alpha|\tau|+\frac{\alpha^2\tau^2}{3}\right)$$

is differentiable twice, since its spectral density is missing

$$S_X(\omega) = \frac{8\alpha^5 d^2}{3\pi(\alpha^2+\omega^2)^3}$$

and

$$\int_{-\infty}^{\infty} \omega^4 S_X(\omega) < \infty$$

PROBLEM 8.7

The initial imperfection $Y_0(x)$ of an infinite beam is a weakly stationary, band-limited random function of the axial coordinate x. Find the spectral density of $d^2Y_0(x)/dx^2$.

SOLUTION 8.7

$$S_{Yo}(\kappa) = \begin{cases} \text{So}, & \text{for } |\kappa| < \kappa_1 \\ 0, & \text{otherwise} \end{cases}$$

The spectral density of $Y_{o''}$ is then

$$S_{Y_o''}(\kappa) = \kappa^4 S_Y(\kappa) = \begin{cases} \kappa^4\,\text{So}, & \text{for } |\kappa| < \kappa_1 \\ 0, & \text{otherwise} \end{cases}$$

In order to find the appropriate autocorrelation functions, we use the Wiener-Khintchine relationship

$$R_{Yo}(\xi) = \int_{-\infty}^{\infty} S_{Yo}(\kappa)e^{i\kappa\xi}d\kappa$$

where $\xi = x_2 - x_1$ is the difference between the observation points along the axis of the beam:

$$R_{Y_o}(\xi) = \int_{-\kappa_1}^{\kappa_1} So\, e^{i\kappa\xi} d\kappa = \frac{S_o}{i\xi} e^{i\kappa\xi} \Big]_{-\kappa_1}^{\kappa_1}$$

$$= \frac{S_o}{i\xi}(e^{i\kappa_1\xi} - e^{-i\kappa_1\xi}) = \frac{So}{i\xi} \cdot 2i \cdot \sin\kappa_1\xi = 2S_o \frac{\sin\kappa_1\xi}{\xi}$$

Now, for $R_{Y''}(\xi)$ we get

$$R_{Y''}(\xi) = \int_{-\kappa_1}^{\kappa_1} \kappa^4 e^{i\kappa\xi} d\kappa = 2\int_0^{\kappa_1} \kappa^4 \cos\kappa\xi d\kappa$$

PROBLEM 8.8

Show that the spectral density of the sum of a pair of independent random functions equals the sum of their spectral densities.

SOLUTION 8.8

The autocorrelation function $R_Z(\tau)$ of the sum of two random functions $X(t)$ and $Y(t)$ is

$$R_Z(\tau) = R_x(\tau) + R_y(\tau) + R_{xy}(\tau) + R_{yx}(\tau)$$

If X and Y are independent,

$$R_{XY}(\tau) = R_{Yx}(\tau) = 0$$

we get

$$R_Z(\tau) = R_x(\tau) + R_Y(\tau)$$

Applying the Fourier transform we get

$$\frac{1}{2\pi}\int_{-\infty}^{\infty} R_Z(\tau)e^{-i\omega\tau} d\tau = \frac{1}{2\pi}\int_{-\infty}^{\infty} R_X(\tau)e^{-i\omega\tau} d\tau + \frac{1}{2\pi}\int_{-\infty}^{\infty} R_Y(\tau)e^{i\omega\tau} d\tau$$

and

$$S_Z(\omega) = S_X(\omega) + S_Y(\omega)$$

instead of the general formulas for correlated X and Y

$$S_Z(\omega) = S_X(\omega) + S_Y(\omega) + S_{XY}(\omega) + S_{YX}(\omega)$$

$$= S_X(\omega) + S_Y(\omega) + 2Re[S_{XY}(\omega)]$$

PROBLEM 8.9

Use the non-negativeness property of the spectral density to determine the admissible values of the parameters α and β in the autocorrelation function

$$R_X(\tau) = d^2 e^{-\alpha|\tau|}\left(\cosh\beta\tau + \frac{\alpha}{\beta}\sinh\beta|\tau|\right)$$

Check whether $X(t)$ is differentiable.

SOLUTION 8.9

The simple check shows that

$$R_X(0) = d^2 > 0$$

and

$$R_X(\tau) = R_X(-\tau)$$

as well as $R_X(\tau)$ does not exceed its value at the origin. Let us check whether the spectral density, associated with this autocorrelation function possesses the property of the nonnegativeness:

$$\begin{aligned}
S_X(\omega) &= \frac{1}{2\pi}\int_{-\infty}^{\infty} R_X(\tau)e^{-i\omega\tau}\,d\tau \\
&= \frac{1}{2\pi}Re\left\{\left(1+\frac{\alpha}{\beta}\right)\int_0^{\infty} e^{-(\alpha-\beta+i\omega)\tau}\,d\tau\right. \\
&\quad \left. + \left(1-\frac{\alpha}{\beta}\right)\int_0^{\infty} e^{-(\alpha+\beta+i\omega)\tau}\,d\tau\right\} \\
&= \frac{1}{2\pi\beta}Re\left\{\frac{\beta+\alpha}{\alpha-\beta+i\omega} + \frac{\beta-\alpha}{\alpha+\beta+i\omega}\right\} \\
&= \frac{\alpha^2-\beta^2}{2\pi\beta}\left\{\frac{1}{\left(\alpha-\beta^2+\omega\right)^2} - \frac{1}{(\alpha+\beta)^2+\omega^2}\right\} \\
&= \frac{\alpha^2-\beta^2}{\pi}\frac{2\alpha}{\left[(\alpha-\beta)^2+\omega^2\right]\left[(\alpha+\beta)^2+\omega\right]}
\end{aligned}$$

This expression is positive only if $\alpha > \beta$. In case of $\alpha + \beta$, $S_X(\omega) + \delta(\omega)$.

PROBLEM 8.10

A beam, simply supported at its ends, is subjected to a distributed force Q, a random variable with given probability density function. Using the relations

$$\frac{dV_y(x)}{dx} = -Q(x) \quad \frac{dM_z(x)}{dx} = -V_y(x)$$

Show that the shear force $V_Y(x)$ and bending moment $M_z(x)$ are the random functions of x. Find

$$E[V_y(x)] \quad E[M_z(x)] \quad R_{V_y}(x_1, x_2) \quad R_{M_z}(x_1, x_2)$$

and the first-order probability densities

$$f_{V_y}(v_y; x) \quad f_{M_z}(m_z; x).$$

SOLUTION 8.10

Since operators involved are linear ones we get

$$\frac{d}{dx}E[V_Y(x)] = -E[Q(x)], \quad \frac{d}{dx}E[M_Z(x)] = -E[V_y(x)]$$

and

$$E[V_y(x)] = -\int_0^x E[Q(z)]dz + C$$

$$E[M_z(x)] = -\int_0^x E[V_y(t)]dt + D$$

where C and D are constants of integration, found from the boundary conditions.

PROBLEM 8.11

A cantilever is subjected to a distributed force $Q_y(x)$ with the zero mean and autocorrelation function:

$$R_{Q_y}(x_1, x_2) = e^{-\alpha|x_2 - x_1|}$$

Using Eq. (8.8), verify that

$$\begin{aligned} R_{M_z}(x_1, x_2) = \frac{1}{\alpha^4}[e^{-\alpha|x_1 - x_2|} - e^{-\alpha x_1} - e^{-\alpha x_2} - \alpha x_1 e^{-\alpha x_2} \\ - \alpha x_2 e^{-\alpha x_1} + 1 + \alpha|x_1 - x_2| - \alpha x_1 x_2 + \frac{\alpha^3}{6}|x_1 - x_2|^3 \\ - \frac{\alpha^3}{6}(x_1^3 - 3x_1^2 x_2 - 3x_1 x_2^2 - 3x_1 x_2^2 + x_2^3)] \end{aligned} \quad (8.164)$$

Show that for $x \geq 1/\alpha$ the variance is

$$\mathrm{Var}\left[M_z(x)\right] = \frac{1}{\alpha^4}\left(2 - \alpha^2 x^2 + \frac{2}{3}\alpha^3 x^3\right) \tag{8.165}$$

Equations (8.164) and (8.165) are due to Rzhanitsyn.

SOLUTION 8.11

Taking into account equations

$$\frac{dV_y(x)}{dx} = -Q_y(x), \quad \frac{dM_z(x)}{dx} = -V_y(x)$$

we get

$$\frac{d^2 M_z(x)}{dx^2} = Q_y(x)$$

Now for the autocorrelation function of $M_z(x)$, namely for $E[M_z(x_1)M_z(x_2)]$ (means are zero) we get

$$\frac{\partial^4 R_{M_z}(x_1 x_2)}{\partial x_1^2 \partial x_2^2} = R_{Q_y}(x_1, x_2) \tag{1}$$

We associate the origin of the coordinates with the cantilever's tip. Since $M_z(0) = 0$, we should get the following boundary conditions for the autocorrelation function of $M_z(x)$:

$$R_{M_z}(0, x_2) = R_{M_z}(x_1, 0) = 0 \tag{2}$$

In addition, at the tip

$$V_y(x) = \frac{dM_z(x)}{dx} = 0$$

Therefore

$$R_{Q_y}(0, x_2) = R_{Q_y}(x_1, 0) = 0 \tag{3}$$

So that we get boundary conditions:

$$R_{M_z}(0, x_2) = R_{M_z}(x_1, 0) = \frac{\partial^2 R_{M_Z}}{\partial x_1 \partial x_2}(0_1 x_2) = \frac{\partial^2 R_{M_z}}{\partial x_1 \partial x_2^1}(x, 0) = 0 \tag{4}$$

The solution of the differential equation (1) can be put as the sum of the complementary and particular solutions. Let us find the latter one (denoted by the asterisk):

$$R_{M_z^*}(x_1 x_2) = \iiiint R_{Q_y}(x_1, x_2)\, dx_1 dx_2 dx_1' dx_2' \tag{5}$$

In order to evaluate this integral we introduce new coordinates:

$$t = x_1 - x_2$$

$$s = x_1 + x_2$$

Eq. (1) becomes then

$$\left(\frac{\partial^2}{\partial s^2} - \frac{\partial^2}{\partial t^2}\left(\frac{\partial^2 R_{M_z}(x_1, x_2)}{\partial s^2} - \frac{\partial^2 R_{M_z}(x_1, x_2)}{\partial t^2}\right)\right) = R_{Q_y}(t)$$

since Q_z is a stationary function. The particular solution is sought as

$$R^*_{MZ}(x_1, x_2) = M^*(t) \tag{6}$$

so that

$$\frac{d^4 M^*(t)}{dt^4} = R_{Q_y}(t) = e^{-\alpha|t|}$$

For $t > 0$ we get

$$\frac{d^3 M^*(t)}{dt^3} = \frac{1}{\alpha}(1 - e^{-\alpha t})$$

$$\frac{d^2 M^*(t)}{dt^2} = \frac{1}{\alpha^2}(\alpha t + e^{-\alpha t} - 1)$$

$$\frac{dM^*}{dt} = \frac{1}{\alpha^3}\left(\frac{\alpha^2 t^2}{2} - e^{-\alpha t} + 1\right)$$

$$M^*(t) = \frac{1}{\alpha^4}(e^{-\alpha t} + \frac{1}{6}\alpha^3 t^3 - \frac{1}{2}\alpha^2 t^2 + \alpha t - 1)$$

For negative t we should change signs of t which are appearing in odd powers. Finally,

$$M^*(t) = \frac{1}{\alpha^4}\left(e^{-\alpha|t|} + \frac{1}{6}\alpha^3|t|^3 - \frac{1}{2}\alpha|t| - 1\right)$$

or

$$R_{M_z^*}(x_1, x_2) = \frac{1}{\alpha^4}\left[e^{-\alpha|x_1-x_2|} + \frac{\alpha}{6}|x_1 - x_2|^3 - \frac{d^2}{2}(x_1 - x_2)^2 + \alpha|x_1 - x_2| - 1\right] \tag{7}$$

The complimentary solution of Eq (1) is

$$f_1(x_1) + f_1(x_2) + x_2 f_3(x_1) + x_1 f_3(x_2)$$

The solution has been in this form since the boundary conditions are symmetric with respect to the bisector of the central coordinate angle. So is the particular solution. Therefore, this type of symmetry should be possessed by the complimentary solution. Thus the general solution reads:

$$R_{M_z}(x_1, x_2) = f_1(x_1) + f_1(x_2) + x_1 f_3(x_2) + x_2 f_3(x_2) + R_{M_z^*}(x_1 - x_2) \tag{8}$$

This yields

$$\frac{\partial^2 R_{M_z}(x_1, x_2)}{\partial x_1 \partial x_2} = f_3'(x_1) + f_3'(x_2) + \frac{\partial^2}{\partial x, \partial x_2} R_{M_z^*}(x_1 - x_2) \tag{9}$$

Satisfying Eqs. (4) we get two functional equations

$$f_1(0) + f_1(x_2) + x_2 f_3(0) + R_{M_z^*}(-x_2) = 0 \tag{10}$$

$$f_3'(0) + f_3'(x_2) + \frac{\partial^2}{\partial x_1 \partial x_2} R_{M_z^*}(-x_2) = 0 \tag{11}$$

But (see Eq. 7)

$$R_{M_z^*}(-x_2) = \frac{1}{\alpha^4}\left(e^{-\alpha x_2} + \frac{\alpha^3 x_2^3}{6} - \frac{\alpha^2 x_2^2}{2} + \alpha x_2 - 1\right) \tag{12}$$

$$\frac{\partial^2 R_{M_z^*}}{\partial x_1 \partial x_2} = -\frac{1}{\alpha^2}(e^{-\alpha|x_1-x_2|} + \alpha|x_1 - x_2| - 1)$$

$$\frac{\partial^2 R_{M_z^*}}{\partial x_1 \partial x_2}(-x_2) = -\frac{1}{\alpha^2}(e^{-\alpha x_2} + \alpha x_2 - 1) \tag{13}$$

Substitution of Eqs (12) and (13) into (11) and (11) yields in:

$$f_1(0) + f_1(x_2) + x_2 f_3(0) + \frac{1}{\alpha^4}\left(e^{-\alpha x_2} + \frac{\alpha^3 x_2^3}{6} - \frac{\alpha^2 x_2^2}{2} + \alpha x_2 - 1\right) = 0 \tag{14}$$

$$f_3^{'}(0) + f_3^{'}(x_2) - \tfrac{1}{\alpha^4}(e^{-\alpha x_2} + \alpha x_2 - 1) = 0 \tag{15}$$

But, via Eq. (14) and (15) we get

$$2f_3^{'}(0) = 0;\ f_3^{'}(x) = \frac{1}{\alpha^3}(e^{-\alpha x} + \alpha x_2 - 1) \tag{16}$$

$$f_3(x) = \frac{1}{\alpha^3}\left(-e^{-\alpha x_2} + \frac{\alpha^2 x_2}{2} - \alpha x_2 + C\right) \tag{17}$$

The function $f_1(x)$ is obtainable from Eq. (14):

$$f_1(x_2) = \frac{1}{\alpha^4}\left(-e^{-\alpha x_1} + \frac{\alpha^3 x_2^3}{2} - \frac{\alpha^2 x_2^2}{2} - 1 - \alpha x_2 + C\right)$$

Substitution into (13) yields

$$\begin{aligned} R_{M_z}(x_1, x_2) = \frac{1}{\alpha^4}\Bigg[& -e^{-\alpha x_1} - e^{-\alpha x_2} - \frac{\alpha^3 x_1^3}{6} - \frac{\alpha^3 x_2^3}{6} - \frac{\alpha^2 x_1^2}{2} \\ & + \frac{\alpha^2 x_2^2}{2} + 2 - \alpha x_1 e^{-\alpha x_2} - \alpha x_2 e^{-\alpha x} + \frac{\alpha^3 x_2^2 x_1}{2} + \frac{\alpha^3 x_1^2 x_2}{2} \\ & - 2\alpha^2 x_1 x_2 + e^{-\alpha|x_1 - x_2|} + \frac{\alpha^3}{6}|x_1 - x_2|^3 \\ & - \frac{\alpha^2}{2}(x_1 - x_2)^2 + \alpha|x_1 - x_2| - 1\Bigg] \end{aligned} \tag{18}$$

which in turn yields an expression in the Problem.

When $x_1 = x_2 = x$ we get

$$E(M_z^2(x)) = \frac{1}{\alpha^4}\left(-2e^{-\alpha x} + \frac{2}{3}\alpha^3 x^3 - \alpha^2 x^2 - 2\alpha x e^{-\alpha x} + 2\right)$$

For $x \gg 1/2$ we get

$$E(M_z^2(x)) = \frac{1}{\alpha^4}\left(\frac{2}{3}\alpha^3 x^3 - \alpha^2 x^2 + 2\right)$$

Exactly the same expressions are obtainable by using the integral representation

$$M_z(x) = \int_0^x Q(t)(x - t)dt$$

therefore

$$R_{M_z}(x_1, x_2) = \int_0^{x_1} \int_0^{x_2} R_{Q_y}(t_1, t_2)(x_1 - t_1)(x_2 - t_2)dt_1 dt_2$$

The autocorrelation function

$$R_{Q_y}(t_1 t_2) = e^{-\alpha|t_1 - t_2|}$$

has different analytical expressions for region $t_1 > t_2$ and the region $t_1 < t_2$ (see the Figure):

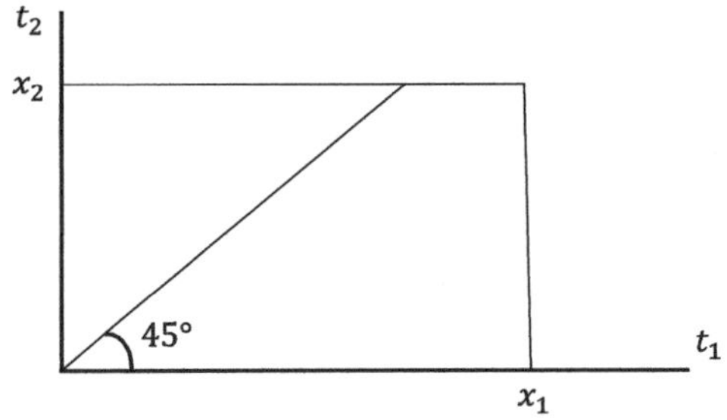

Final results coincides with the Eq. 18.

PROBLEM 8.12

A normal random function $X(t)$ with zero mean has an autocorrelation function as per Eq. (8.106). Find the probability of $X(t) < x_0, x_0$ being a deterministic positive constant.

SOLUTION 8.12

$X(t)$ is a normal random variable (for fixed t) with zero mean and the variance equal d^2, or $N(0, d^2)$ therefore

$$P(X(t) < x_0) = \frac{1}{2} + \text{erf}\left(\frac{x_0}{\sigma_X}\right) = \frac{1}{2} + \text{erf}\left(\frac{x_0}{d}\right)$$

<u>**Note to lecturer**</u>

In order to use "fully" the information on the autocorrelation function, one can ask what is the probability that the first derivative $X'(t)$ will not exceed x_0, which is a determination constant. The autocorrelation function $RX'(\tau)$ is given by (see Eq. 8.118)

$$R_x^{'}(\tau) = -\frac{d^2 R_x(\tau)}{d\tau^2}$$

Now

$$\frac{R_X'(\tau)}{d^2} = \frac{d^2}{d\tau^2} e^{-\alpha|\tau|}(\cos\beta\tau + \frac{\alpha}{\beta}\sin\beta|\tau|)$$

$$= -\frac{d}{d\tau}\left[-\alpha e^{-\alpha|\tau|}\left(\cos\beta\tau + \frac{\alpha}{\beta}\sin\beta|\tau|\right)\frac{d|\tau|}{d\tau}\right.$$

$$\left.+ e^{-\alpha|\tau|}\left(-\beta\sin\beta\tau + \alpha\cos\beta|\tau|\cdot\frac{d|\tau|}{d\tau}\right)\right]$$

$$= \frac{\alpha^2+\beta^2}{\beta}\frac{d}{d\tau}\left[e^{-\alpha|\tau|}\sin\beta\tau\right]$$

$$= \frac{\alpha^2+\beta^2}{\beta}\left[-\alpha e^{-\alpha|\tau|}\sin\beta\tau\frac{d|\tau|}{d\tau} + \beta\cos\beta\tau e^{-\alpha|\tau|}\right.$$

$$= (\alpha^2+\beta^2)e^{-\alpha|\tau|}\left[\cos\beta\tau - \frac{\alpha}{\beta}\sin\beta|\tau|\right]$$

Therefore, for fixed t, $V = X'(t)$ is a normal random variable with the zero mean and variance $d^2(\alpha^2+\beta^2)$. Hence

$$P(X^{'}(t) < \dot{x}_0) = \frac{1}{2} + \text{erf}\left(\frac{\dot{x}_0}{d\sqrt{\alpha^2+\beta^2}}\right)$$

chapter 9

Random Vibration of Discrete Systems

PROBLEM 9.1

A single-degree-of-freedom system (Eq. 9.48) is subjected to a random load (see figure) with exponential autocorrelation function $R_F(\tau) = \pi \alpha S_0 e^{-\alpha|\tau|}$.

(a) Find the mean-square value of the displacement, velocity, and acceleration.
(b) Check that for $\alpha \to \infty$, the values obtained in (a) coincide with those of a system under ideal white-noise excitation.

SOLUTION 9.1

The equation of motions reads

$$m\ddot{x} + c\dot{x} + (1+\varepsilon)kx = F(t)$$

or

$$\ddot{x} + 2\zeta\omega_0^2 x + \omega_0^2 x = \frac{1}{m}F(t)$$

with

$$\omega_0 = \sqrt{(1+\varepsilon)\frac{k}{m}},\ \zeta = \frac{c}{2m\omega_0}$$

Now

$$H(\omega) = \frac{1}{m[i\omega]^2 + 2\zeta(i\omega)\omega_0 + \omega_0^2]}$$

with spectral density

$$S_X(\omega) = \frac{1}{m^2} \frac{S_F(\omega)}{\left[\left(\omega_0^2 - \omega^2\right)^2 + 4\zeta^2\omega_0^2\omega^2\right]}$$

$$= \frac{S_0}{m^2} \frac{\alpha^2}{\left[\left(\omega_0^2 - \omega^2\right)^2 + 4\zeta^2\omega_0^2\omega^2\right]} \frac{1}{\alpha^2 + \omega^2}$$

Now

$$\frac{\alpha^2}{[(\omega_0^2 - \omega^2)^2 + 4\zeta^2\omega_0^2\omega^2](\alpha^2 + \omega^2)} = \frac{A(\omega)}{\left(\omega_0^2 - \omega^2\right)^2 + 4\zeta^2\omega_0^2\omega^2} + \frac{B(\omega)}{\alpha^2 + \omega^2}$$

with

$$A(\omega) = a + b\omega + c\omega^2 + d\omega^3$$

$$b(\omega) = e + f\omega$$

Comparison shows

$$a = \frac{\alpha^2}{\alpha^2 - \frac{1}{\beta^2 - 2\omega_0^2 - \alpha^2}}$$

$$e = -c = -\frac{\alpha^2}{\alpha^2\left(\beta_0^2 - 2\omega^2 - \alpha^2\right) - 1}$$

$$b = d = f = 0$$

and

$$S_X(\omega) = \frac{S_0}{m^2}\left[\frac{a + c\omega^2}{\left(\omega_0^2 - \omega^2\right)^2 + \beta^2\omega^2} + \frac{e}{\alpha^2 + \omega^2}\right]$$

$$E\left(X^2\right) = \frac{S_0}{m^2}\int_{-\infty}^{\infty} \frac{\left(a + c\omega^2\right) d\omega}{\left(\omega_0^2 - \omega^2\right)^2 + \beta^2\omega^2} + \frac{S_0}{m^2}\int_{-\infty}^{\infty} \frac{e d\omega}{\alpha^2 + \omega^2}$$

The first integral equals (see pp. 475–476):

$$\int_{-\infty}^{\infty} \frac{\left(a + c\omega^2\right) d\omega}{\left(\omega_0^2 - \omega^2\right)^2 + \beta^2\omega^2} = \frac{(a + c\omega_0^2)\pi}{2\zeta\omega_0^3}$$

whereas the second is (see, e.g., p. 477)

$$\int_{-\infty}^{\infty} \frac{ed\omega}{\alpha^2+\omega^2} = \frac{\pi}{\alpha}$$

and

$$E\left(X^2\right) = \frac{S_0\pi}{m^2}\left[\frac{a+c\omega_0^2}{2\zeta\omega_0^3} + \frac{1}{\alpha}\right]$$

When $\alpha \to \infty$, $c \to 0$, $w \to 0$ and $a \to 1$, and

$$E\left(X^2\right) = \frac{S_0\pi}{2\zeta\omega_0^3 m^2}$$

i.e. we get response of a single-degree-of-freedom system to the white noise excitation.

The solution of the problem can be furnished in a different way, by making use of Appendix C (p. 472). Then

$$E\left(X^2\right) = \frac{S_0\alpha^2}{m^2}\int_{-\infty}^{\infty} \frac{d\omega}{\left|\left[(i\omega)^2+2\zeta\omega_0(i\omega)+\omega_0^2\right][i\omega+\alpha]\right|^2}$$

i.e. in this case

$$L_3(i\omega)^3 + (i\omega)^2(2\zeta\omega_0+\alpha) + (i\omega)\left(\omega_0^2+2\zeta\omega_0\alpha\right) + \alpha\omega_0^2$$

or

$$a_0 = 1, \quad a_1 = 2\zeta\omega_0+\alpha, \quad a_2 = \omega_0^2+2\zeta\omega_0\alpha, \quad a_3 = \alpha\omega_0^2$$

$$b_0 = 0, \quad b_1 = 0, \quad b_2 = 1$$

formula (C.5) yields then

$$I_3 = \frac{\pi}{a_0(a_0a_3-a_1a_2)}\left(-\frac{a_0a_1}{a_3}b_2\right)$$

and

$$a_0a_3 - a_1a_2 = -\omega_0(2\zeta\omega_0^2+4\zeta^2\omega_0\alpha+2\zeta\alpha^2)$$

resulting in

$$E\left(X^2\right) = \frac{S_0\alpha^2}{m^2}I_3 = \frac{S_0\alpha^2}{m^2}\frac{\pi}{\omega_0(2\zeta\omega_0^2+4\zeta^2\omega_0\alpha+2\zeta\alpha^2)}\frac{2\zeta\omega_0+\alpha}{\alpha\omega_0^2}$$

$$= \frac{S_0\pi}{m^2}\frac{(2\zeta\omega_0+\alpha)\alpha}{\omega_0^3(2\zeta\omega_0^2+4\zeta^2\omega_0\alpha+2\zeta\alpha^2)}$$

When $\alpha \to \infty$, we get

$$\frac{2\zeta\omega_0\alpha + \alpha^2}{2\zeta\omega_0^2 + 4\zeta^2\omega_0\alpha + 2\zeta\alpha^2} + \frac{1}{2\zeta}$$

and

$$E\left(X^2\right) = \frac{S_0\pi}{2\zeta\omega_0^3 m^2}$$

which coincides with the response of the single-degree-of-freedom system under white noise excitation.

For the mean-square velocity we have:

$$E\left(\dot{X}^2\right) = \frac{S_0\alpha^2}{m^2}\int_{-\infty}^{\infty}\frac{\omega^2 d\omega}{|\left[(i\omega)^2 + 2\zeta\omega_0(i\omega) + \omega_0^2\right][i\omega + \alpha]|^2}$$

with the same "a"s but with

$$b_0 = 0, \quad b_1 = -1, \quad b_2 = 0$$

Therefore, Equation 6.5 yields in

$$I_3 = \frac{\pi}{a_0(a_0a_3 - a_1a_2)}a_0b_1$$

and

$$E\left(\dot{X}^2\right) = \frac{1}{\omega_0(2\zeta\omega_0^2 + 4\zeta^2\omega_0\alpha + 2\zeta\alpha^2)}$$

when $\alpha \to \infty$, we get

$$E\left(\dot{X}^2\right) = \frac{S_0\pi}{2\zeta\omega_0 m^2}$$

i.e. coincides with the result of the single-degree-of-freedom system. Now, for the mean-square acceleration:

$$E\left(\ddot{X}^2\right) = \frac{S_0\pi}{m^2}\int_{-\infty}^{\infty}\frac{\omega^4 d\omega}{|\left[(i\omega)^2 + 2\zeta\omega_0(i\omega) + \omega_0^2\right][i\omega + \alpha]|^2}$$

we have the same "a"s but with

$$b_0 = 1, \quad b_1 = 0, \quad b_2 = 0$$

and (see Eq. C.5, p. 972)

$$I_3 = \frac{\pi}{a_0(a_0a_3 - a_1a_2)}(-a_2b_0);$$

Therefore

$$E\left(\ddot{X}^2\right) = \frac{S_0\alpha^2\pi}{m^2}\frac{\omega_0^2 + 2\zeta\omega_0\alpha}{\omega_0(2\zeta\omega_0^2 + 4\zeta\omega_0\alpha + 2\zeta\alpha^2)}$$

this expression is finite for finite α and tends to infinity for $\alpha \to \infty$. This agrees with the result of Example 9.5 (see conclusion on p. 336 about the $E(\ddot{X}^2)$).

PROBLEM 9.2

A cantilever with a concentrated mass attached to its tip is subjected to random loading with autocorrelation function in the form of ideal white noise with intensity S_0. The cantilever itself is massless. Geometric dimensions are indicated in the accompanying figure. Find the mean-square displacement and the mean-square velocity.

SOLUTION 9.2

The deflection of the cantilever under the loading P at its tip is $W = \frac{Pl^3}{3EI}$ which is equivalent, if we replace the beam by a spring with stiffness k, to

$$k = \frac{3EI}{l^3}$$

Equation of motion of this single-degree-of-freedom system is

$$\ddot{x} + \frac{c}{m}\dot{x} + \frac{3EI}{l^3 m}x = \frac{1}{m}F(t)$$

The natural frequency is

$$\omega_0 = \sqrt{\frac{k}{m}} = \sqrt{\frac{3EI}{l^3 m}}$$

The damping coefficient is

$$\zeta = \frac{c}{2m\omega_0}$$

therefore, see Equation 9.68

$$E\left(X^2\right) = \frac{S_0\pi}{2\zeta\omega_0^3 m^2} = \frac{S_0\pi}{2}\frac{2m\omega_0}{c}\frac{1}{\omega_0^3 m^2}$$

$$= \frac{S_0\pi}{c\omega_0^2 m} = \frac{S_0\pi\, al^3}{3EIc}$$

i.e. is independent of the mass. The mean square velocity

$$E\left(\dot{X}^2\right) = \frac{s_0\pi}{mc}$$

i.e. is independent of the stiffness.

PROBLEM 9.3

A road vehicle travels with uniform velocity v on a rough surface (see figure) and in the process is subjected to a time-varying displacement excitation. The roughness of the profile is a random function of the coordinate x, hence both the excitation and response of the vehicle are random. Derive an expression for the response mean-square value if the autocorrelation function of the profile is $R_Y(x_1, x_2) = d^2 \exp -\alpha|x_2 - x_1|$.

SOLUTION 9.3

The equation of motion reads

$$m\ddot{x} = -k(x - y) - c(\dot{x} - \dot{y})$$

or

$$m\ddot{x} + c\dot{x} + kx = c\dot{y} + ky$$

Now

$$R_Y\left(x_1, x_2\right) = d^2 \exp\left[-\alpha\left|x_2 - x_1\right|\right]$$

with

$$x_2 = vt_2, \quad x_1 = vt_1$$

resulting in

$$R_Y\left(x_1, x_2\right) = d^2 \exp\left[-\alpha v\left|t_2 - t_1\right|\right] = d^2 \exp[-\exp[-\alpha v\left|\tau\right|] \equiv R_Y(\tau)$$

where $\tau = t_2 - t_1$. Denoting $\gamma = \alpha v$, we get

$$R_Y\left(\tau\right) = d^2 \exp[-\gamma\left|\tau\right|]$$

with spectral density (see Equation 8.90, p. 297, or Appendix D, p. 477)

$$S_Y\left(\omega\right) = \frac{d^2\gamma}{\pi}\frac{1}{\gamma^2 + \omega^2} = \frac{d^2\alpha v}{\pi}\frac{1}{(\alpha v)^2 + \omega^2}$$

Via Equation (9.35) we get

$$S_X(\omega) = |H(\omega)|^2 S_Y(\omega)$$
$$= \frac{1}{m^2}\frac{1}{(\omega_0^2-\omega^2)^2+4\zeta^2\omega_0^2\omega^2}\frac{d^2\alpha v}{\pi}\frac{1}{(\alpha v)^2+\omega^2}$$

so that for the response mean-square we obtain an expression

$$E\left(X^2\right) = \int_{-\infty}^{\infty} S_X(\omega)\,d\omega$$
$$= \frac{d^2\alpha v}{\pi m^2}\int_{-\infty}^{\infty}\frac{1}{(\omega_0^2-\omega^2)^2+4\zeta^2\omega_0^2\omega^2}\frac{1}{(\alpha v)^2+\omega^2}d\omega$$

which is analogous to that obtained in Problem 9.1, with attendant exact expression available by repeating the steps outlined in the solution of Problem 9.1. For approximate solution we resort to the Laplace's asymptotic method, as described in pp. 336–338, yielding

$$E(X^2) \approx \frac{S_F(\omega_0)\pi}{2\zeta\omega_0^3 m^2}$$

where $S_F(\omega)$

$$S_F(\omega) = \frac{d^2\alpha v}{\pi[(\alpha v)^2+\omega^2]}$$

and

$$E\left(X^2\right) = \frac{d^2\alpha v}{\pi\left[(\alpha v)^2+\omega_0^2\right]}\frac{\pi}{2\zeta\omega_0^3 m^2}$$

But

$$\zeta = \frac{c}{2m\omega_0}, \quad \omega_0^2 = \frac{k}{m}$$

so that we finally get

$$E(X^2) \approx \frac{d^2\alpha v}{ck[(\alpha v)^2+(k/m)^2]}$$

PROBLEM 9.4

Consider a nonsymmetric two-degree-of-freedom system under ideal white-noise excitation $F_1(t)$.

(a) Verify that the error incurred by disregarding the cross-correlations in evaluating $E(|X_1|^2)$ reaches 40% when ε tends to zero.

(b) Verify that at $\varepsilon = 1$ the cross-correlation terms can still be omitted but $E(|X_1|^2)$ differs from $E(|X_2|^2)$, unlike the example of a symmetric system considered in Sec. 9.4.

SOLUTION 9.4

The equation of motion reads:

$$m\ddot{x}_1 + (1+\varepsilon)\,c\dot{x} - \varepsilon c\dot{x}_2 + (1+\varepsilon)\,kx_1 - \varepsilon kx_2 = F_1(t)$$

$$(1+\varepsilon)\,m\ddot{x}_2 + (1+\varepsilon)\,c\dot{x}_2 - \varepsilon c\dot{x}_1 + (1+\varepsilon)\,kx_2 - \varepsilon k\bar{x}_1 = 0$$

Matrices M, K and C are

$$[M] = \begin{bmatrix} m & 0 \\ 0 & m(1+\varepsilon) \end{bmatrix}$$

$$[K] = \begin{bmatrix} k(1+\varepsilon) & -k\varepsilon \\ -k\varepsilon & k(1+\varepsilon) \end{bmatrix}$$

$$[C] = \begin{bmatrix} c(1+\varepsilon) & -c\varepsilon \\ -c\varepsilon & c(1+\varepsilon) \end{bmatrix}$$

Natural frequencies are found from

$$\begin{vmatrix} k\,(1+\varepsilon) - m\omega^2 & -k\varepsilon \\ -k\varepsilon & k\,(1+\varepsilon) - m(1+\varepsilon)\omega^2 \end{vmatrix}$$

or

$$(1+\varepsilon)\,m^2\omega^4 - km\,(1+\varepsilon)\,(2+\varepsilon)\,\omega^2 + k^2\,(1+2\varepsilon) = 0$$

with

$$\omega_{1,2}^2 = \frac{k}{m}\left[\frac{(1+\varepsilon)\,(2+\varepsilon) \mp \sqrt{5\varepsilon^2 + 6\varepsilon^3 + \varepsilon^4}}{2\,(1+\varepsilon)}\right]$$

For the first mode we find

$$\{k\,(1+\varepsilon) - m\omega_1^2 - k\varepsilon\}\begin{Bmatrix} y_1^{(1)} \\ y_2^{(1)} \end{Bmatrix}$$

and with $y_1^{(1)} = \lambda_1$ we get

$$y_2^{(1)} = \lambda_1 \frac{1+\varepsilon - [(1+\varepsilon)(2+\varepsilon) - \sqrt{5\varepsilon^2 + 6\varepsilon^3 + \varepsilon^4}]/2(1+\varepsilon)}{\varepsilon} \equiv \alpha\lambda_1$$

Accordingly, for the second mode, $y_2^{(1)} = \lambda_2$ and

$$y_2^{(2)} = \lambda_2 \frac{1+\varepsilon - [(1+\varepsilon)(2+\varepsilon) - \sqrt{5\varepsilon^2 + 6\varepsilon^3 + \varepsilon^4}]/2(1+\varepsilon)}{\varepsilon} \equiv \beta\lambda_1$$

We normalize natural modes and construct the modal matrix

$$[v] = \begin{bmatrix} \dfrac{\lambda_1}{\mu_1} & \dfrac{\lambda_2}{\mu_2} \\ \dfrac{\lambda_1}{\mu_1}\alpha & \dfrac{\lambda_2}{\mu_2}\beta \end{bmatrix}$$

with

$$[v]^T [M][V] = [I]$$

we get

$$\frac{\lambda_1^2}{\mu_1^2} m + \frac{\lambda_1^2}{\mu_1^2} \alpha^2 m\,(1+\varepsilon) = 1$$

$$\frac{\lambda_1}{\mu_1} = \frac{1}{\sqrt{m}\sqrt{1+\alpha^2(1+\varepsilon)}}$$

Analogically

$$\frac{\lambda_2}{\mu_2} = \frac{1}{\sqrt{m}\sqrt{1+\beta^2(1+\varepsilon)}}$$

and

$$[v] = \frac{1}{\sqrt{m}} \begin{bmatrix} [1+\alpha^2(1+\varepsilon)]^{-1/2} & [1+\beta^2(1+\varepsilon)]^{-1/2} \\ \alpha[1+\alpha^2(1+\varepsilon)]^{-1/2} & \beta[1+\beta^2(1+\varepsilon)]^{-1/2} \end{bmatrix}$$

Now

$$[S_\Phi(\omega)] = [v]^T [S_F][v]$$

$$\frac{S_0}{m} = \begin{bmatrix} \dfrac{1}{1+\alpha^2(1+\varepsilon)} & \dfrac{1}{\sqrt{1+\alpha^2(1+\varepsilon)}\sqrt{1+\beta^2(1+\varepsilon)}} \\ \dfrac{1}{\sqrt{1+\alpha^2(1+\varepsilon)}\sqrt{1+\beta^2(1+\varepsilon)}} & \dfrac{1}{1+\beta^2(1+\varepsilon)} \end{bmatrix}$$

$$[S] \equiv [II]\,[S_\Phi]\,[II]^*$$

$$= \frac{S_0}{m}\begin{bmatrix} \dfrac{1}{1+\alpha^2(1+\varepsilon)}|H_1|^2 & \dfrac{1}{\sqrt{1+\alpha^2(1+\varepsilon)}\sqrt{1+\beta^2(1+\varepsilon)}}H_1H_2^* \\ \dfrac{1}{\sqrt{1+\alpha^2(1+\varepsilon)}\sqrt{1+\beta^2(1+\varepsilon)}}H_2H_1^+ & \dfrac{1}{1+\beta^2(1+\varepsilon)}|H_2|^2 \end{bmatrix}$$

Therefore the spectral density of the displacements becomes

$$S_{x_1}(\omega) = [v_1]\,[S]\,[v_1]^T$$

$$= \frac{S_0}{M^2}\left\{\frac{1}{[1+\alpha^2(1+\varepsilon)]^2}\,|H_1|^2 + [1+\alpha^2\,(1+\varepsilon)\right.$$

$$\times\,[1+\beta^2(1+\varepsilon)]|H_2|^2$$

$$\left. + 2\frac{\alpha}{[1+\alpha^2\,(1+\varepsilon)]^{3/2}\sqrt{1+\beta^2(1+\varepsilon)}}\mathrm{Re}H_1^*H_2\right\},$$

$$S_{X_2} = \frac{S_0}{m^2}\left\{\frac{\beta^2}{[1+\alpha^2(1+\varepsilon)][1+\beta^2(1+\varepsilon)]}|H_1|^2\right.$$

$$+\,\frac{1}{[1+\beta^2(1+\varepsilon)]^2}|H_2|^2$$

$$\left. + 2\frac{\beta}{\sqrt{1+\alpha^2(1+\varepsilon)]}\,[1+\beta^2\,(1+\varepsilon)]^{3/2}}\mathrm{Re}H_1H_2^*\right\}$$

Let us investigate what happens when $\varepsilon \to 0$:

$$\alpha \to 1, \quad \beta \to -1, \quad \omega_1^2 \to \omega_2^2 \to \frac{k}{m}$$

and

$$S_{X_1}(\omega) = \frac{S_0}{m^2}\left[\frac{1}{4}\,|H_1|^2 + \frac{1}{4}\,|H_2|^2 + \frac{1}{4}2\cdot\mathrm{Re}H_1H_2^*\right]$$

The mean square displacement becomes

$$E(|X_1|^2) = \frac{S_0\pi}{4ck}\,(1+1+2) = \frac{S_0\pi}{ck}$$

For $\varepsilon \to 1$

$$\omega_1^2 = \frac{k}{m}\frac{3-\sqrt{3}}{2} = 0.634\frac{k}{m}; \quad \alpha = 2 - \frac{3-\sqrt{3}}{2} = 1.366$$

$$\omega_2^2 = \frac{k}{m}\frac{3+\sqrt{3}}{2} = 2.366\frac{k}{m}; \quad \beta = 2 - \frac{3+\sqrt{3}}{2} = -0.366$$

$$S_{X_{1,2}} = \frac{S_0}{m^2}\left[\frac{1}{22.4}|H_1|^2 + \frac{1.866}{6}|H_2|^2 \mp 2\cdot\frac{1.366}{11.6}\mathrm{Re}H_1H_2^*\right]$$

The mean square displacements become

$$E(|X_1|^2) = \frac{\pi S_0}{ck}\left[\frac{1}{22.4\cdot 0.634} + \frac{1.866}{6.2\cdot 2.366} + \frac{1.366\cdot 12}{3.1\cdot 39}\right.$$

$$= \frac{\pi S_0}{ck}(0.07 + 0.13 + 0.036)$$

$$E(|X_1|^2) = \frac{\pi S_0}{ck}\left(\frac{0.134\cdot 1}{6\cdot 0.634} + \frac{1}{1.6\cdot 2.366} - \frac{1.366\cdot 12}{11.6\cdot 39}\right)$$

$$= \frac{\pi S_0}{ck}(0.035 + 0.264 - 0.036)$$

Neglecting cross-correlations for $\varepsilon = 1$ yields error of 15.25% for the first mass, and 13.69% for the second mass.

PROBLEM 9.5

The system shown in the figure is subjected to random loading with the autocorrelation function

$$R_F(\tau) = d^2 e^{-\alpha|\tau|}\left(\cos\beta\tau + \frac{\alpha}{\beta}\sin\beta|\tau|\right)$$

The masses are attached to a massless beam simply supported at its ends, with stiffness modulus EI. Find $E(|X_1|^2)$ and $E(|X_2|^2)$.

SOLUTION 9.5

The mass and damping matrices are

$$[m] = m\begin{bmatrix}1 & 0\\ 0 & 1\end{bmatrix}, \quad [c] = c\begin{bmatrix}1 & 0\\ 0 & 1\end{bmatrix}$$

The stiffness matrix can be obtained from the book by Popov E., "Introduction to Mechanics of Solids", pp. 395–397:

$$[K] = k \begin{bmatrix} 1 & 7/8 \\ 7/8 & 1 \end{bmatrix}$$

where

$$k = \frac{234}{4}\frac{EI}{L^3} = \frac{243}{4}\frac{EI}{27l^3} = \frac{9}{4}\frac{EI}{l^3}$$

To find the natural frequencies we solve equation

$$\begin{vmatrix} k - m\omega^2 & (7/8)k \\ (7/8)k & k - m\omega^2 \end{vmatrix} = 0$$

or

$$\left(k - m\omega^2\right)^2 - \frac{49}{64}k^2 = 0$$

$$k - m\omega^2 = \mp\frac{7}{8}\frac{k}{m}$$

with

$$\omega_{1,2}^2 = \frac{k}{m} \mp \frac{7}{8}\frac{k}{m},$$

$$\omega_1^2 = \frac{1}{8}\frac{k}{m}, \quad \omega_2^2 = \frac{15}{8}\frac{k}{m}$$

for the first mode we get

$$\left\{k - \frac{k}{8} \quad \frac{7}{8}k\right\} \begin{Bmatrix} y_1^{(1)} \\ y_2^{(1)} \end{Bmatrix} = 0$$

yielding

$$\left\{y^1\right\} = \lambda_1 \begin{Bmatrix} 1 \\ 1 \end{Bmatrix}, \quad \text{for the second mode} \left\{k - \frac{k}{8} \quad \frac{7}{8}k\right\} \begin{Bmatrix} y_1^{(1)} \\ y_2^{(1)} \end{Bmatrix} = 0$$

$$\text{yielding } y^{(2)} = \lambda_2\{1\}$$

We normalize natural modes and construct the modal matrix

$$[v] = \begin{vmatrix} \dfrac{1}{\mu_1} & \dfrac{\lambda_2}{\mu_2} \\ -\dfrac{\lambda_1}{\mu_1} & \dfrac{\lambda_2}{\mu_2} \end{vmatrix}$$

Requirement

$$[V]^T[M][v] = [I]$$

yields

$$\frac{\lambda_1}{\mu_1} = \frac{\lambda_2}{\mu_2} = \frac{1}{\sqrt{2m}}$$

and

$$[v] = \frac{1}{\sqrt{2m}} \begin{bmatrix} 1 & 1 \\ -1 & 1 \end{bmatrix}$$

$$R_F(\tau) = d^2 e^{-\alpha|\tau|} \left(\cos \beta\tau + \frac{\alpha}{\beta} \sin \beta|\tau| \right)$$

we have

$$S_{F_1}(\omega) = \frac{d^2}{\pi} \frac{(\alpha + \gamma\beta)\left(\alpha^2 + \beta^2\right) + (\alpha - \gamma\beta)\omega^2}{\left(\alpha^2 + \beta^2\right)^2 + 2\left(\alpha^2 - \beta^2\right)\omega^2 + \omega^4}$$

(see p. 477); or with $\gamma = \alpha/\beta$ we get

$$S_{F_1}(\omega) = \frac{2\beta(\alpha^2 + \beta^2)}{\left(\alpha^2 + \beta^2\right)^2 + 2\left(\alpha^2 - \beta^2\right)\omega^2 + \omega^4}$$

The matrix $[S_F(\omega)]$ becomes

$$[S_F(\omega)] = \begin{bmatrix} S_{F_1}(\omega) & 0 \\ 0 & 0 \end{bmatrix}$$

Now

$$\begin{aligned} [S_\Phi(\omega)] &= [V]^T [S_F(\omega)] [v] \\ &= \frac{S_{F_1}(\omega)}{2m} \begin{bmatrix} 1 & -1 \\ 1 & -1 \end{bmatrix} \begin{bmatrix} 1 & 0 \\ 0 & 0 \end{bmatrix} \begin{bmatrix} 1 & 1 \\ -1 & 1 \end{bmatrix} = \frac{S_{F_1}(\omega)}{2m} \begin{bmatrix} 1 & 1 \\ 1 & 1 \end{bmatrix} \end{aligned}$$

So that

$$S_{X_1}(\omega) = [v_1][H(\omega)][S_\Phi(\omega)][H(\omega)]^*[v_1]^T$$

and

$$E(|X_1|^2) = \frac{1}{4m^2} \int_{-\infty}^{\infty} S_{F_1}(\omega)[|H_1|^2 + |H_2|^2 + 2 \operatorname{Re} H_1 H_2^*] d\omega$$

For light damping,

$$1-\frac{\omega_1^2}{\omega_2^2}=1-\frac{1/64}{225/64}=0.996\gg\zeta_1^2,\zeta_2^2$$

(see Equation 9.203, p. 371) and the cross-correlation terms can be neglected.

In addition, using Laplace's asymptotic evaluation of integrals, as discussed on pp. 336–337, we get

$$E(|X_1|^2)\simeq E(|X_2|^2)=\frac{\pi}{4m^2}\left[\frac{S_{F_1}(\omega)}{2\zeta_1\omega_1^3}+\frac{S_{F_1}(\omega_2)}{2\zeta_2\omega_2^3}\right]$$

$$=\frac{d^2}{4kc}\cdot 2\alpha(\alpha^2+\beta^2)\left[\frac{1}{(\alpha^2+\beta^2)^2+2\left(\alpha^2-\beta^2\right)\frac{1}{8}\frac{k}{m}+\frac{1}{64}\frac{k^2}{m^2}}\right.$$

$$\left.+\frac{1}{(\alpha^2+\beta^2)^2+2\left(\alpha^2-\beta^2\right)\frac{15}{8}\frac{k}{m}+\frac{225}{64}\frac{k^2}{m^2}}\right]$$

If

$$\alpha=\beta=A\sqrt{\frac{k}{m}}$$

then

$$E(|X_1|^2)\simeq E(|X_2|^2)=\frac{d^2}{4kc}\cdot 2A\sqrt{\frac{k}{m}}\left(2A^2\frac{k}{m}\right)x$$

$$\times x\left[\frac{m}{\left(4A^2+\frac{1}{64}\right)k^2}+\frac{m}{\left(4A^2+\frac{225}{64}\right)k^2}\right]$$

$$=\frac{2d^2}{kc}\cdot\left(\frac{k}{m}\right)^{\frac{1}{2}+1-2}A^3\left[\frac{1}{4A^2+\frac{1}{64}}+\frac{1}{4A^2+\frac{225}{64}}\right]$$

$$=\frac{2d^2}{c}\frac{\sqrt{M}}{k\sqrt{k}}A_3\left(\frac{1}{4A^2+\frac{1}{64}}+\frac{1}{4A^2+\frac{225}{64}}\right)$$

For $A=\frac{1}{4}$, we get

$$E(|X_1|^2)\simeq E(|X_2|^2)=\frac{2d^2}{c}\frac{\sqrt{m}}{k\sqrt{k}}\left(\frac{1}{17}+\frac{1}{241}\right)=\frac{516}{4097}\frac{d^2\sqrt{m}}{ck\sqrt{k}}=0.126\frac{d^2\sqrt{m}}{ck\sqrt{k}}$$

The error due to the neglect of the second term is $n=6.6\%$.

PROBLEM 9.6

A beam is clamped at its ends, but free sliding is permitted in the axial direction. The beam itself is massless. The stiffness modulus of the beam between the masses is εEI, where ε is a nonnegative parameter. $F_1(t)$ represents an ideal white-noise excitation with intensity S_0.

(a) Examine the dependence of $E(|X_1|^2)$ versus the parameter ε.
(b) Verify that for $\varepsilon \to 0$ the result of Prob. 9.2 is obtained.

SOLUTION 9.6

Consider first the free vibrations: The equations, the displacements under masses m_1 and m_2 are, under forces P_1 and P_2, applied at them:

$$y_1 = \delta_{11}P_1 + \delta_{12}P_2$$

$$y_2 = \delta_{21}P + \delta_{22}P_2$$

where δ_{jk} is a flexibility coefficient, defined as the displacement in the jth location due to the unit force applied at kth location. Now, the forces are, under free vibration:

$$P_i = -m_i\ddot{y}_i$$

and the free vibration equations become

$$\delta_{11}m_1\ddot{y}_1 + \delta_{12}m_2\ddot{y}_2 + y_1 = 0 \tag{1}$$

$$\delta_{21}m_1\ddot{y}_1 + \delta_{22}m_2\ddot{y}_2 + y_2 = 0$$

To find δ_{jk} we have to solve the problem of strength of materials. The beam is clamped at both of its ends, so that the problem is statically undermined to the second degree. Simpler solution will be obtained by the Galerkin method (see Appendix E, pp. 477–482).

We assume the comparison functions to be

$$\psi_1(x) = x^2(x-3l)^2 = x^4 - 6x^3l + 9l^2x^2$$

$$\psi_2(x) = x^4(x-3l)^2 = x^6 - 6x^5l + 9l^2x^4$$

We substitute the approximation expression of the displacement

$$w(x) = A\psi_1 + B\psi_2$$

into the differential equation

$$L(w) = EI\frac{d^4 w}{dx^4} = \delta(x-l)$$

where $\delta(x-l)$ is a Dirac's delta function, l is the coordinate of the force application.

The left-hand side becomes

$$L(\bar{w}) = \begin{cases} \left(A + 9Bl^2 - 30Blx + 15Bx^2\right), & \text{for } 0 \le x \le l \text{ or } 2l \le x \le 3l \\ 24\varepsilon EI\left(A + 9Bl^2 - 30Blx + 15Bx^2\right), & \text{for } l \le x \le 2l \end{cases}$$

with the "error" being

$$\varepsilon(X) = L(\bar{w}) - \delta(x-l)$$

We multiply the "error" by $\psi_j(x)$ and integrate over the entire span, to get two equations for A and B

$$(238 + 329\varepsilon)A + (1812 - 354\varepsilon)l^2 B = \frac{35}{3}\frac{1}{EIl}$$

$$(1 - 38 + 1149\varepsilon)A + (19142 + 541\varepsilon)l^2 B = \frac{35}{2}\frac{1}{EIl}$$

which results in A and B. Final expression for $w(x)$ is

$$w(x) = \frac{7x^2}{6EIl^3(116947\varepsilon^2 + 942388\varepsilon + 534988)}[19296l^4\varepsilon + 295632l^4$$
$$-12864l^3\varepsilon x - 197088l^3 x - 9655l^2\varepsilon x^2 + 20590l^2x^2$$
$$+7866l\varepsilon x^3 + 8172lx^3 - 1311\varepsilon 3x^4 - 1362x^4]$$

So that for $\varepsilon = 1$

$$\delta_{11}(1) = w(l) = \frac{1862}{19683}\frac{l^3}{EI} = 0.0945994\frac{l^3}{EI}$$

instead of exact solution, readily available for this case

$$\delta_{11} = \frac{8}{81}\frac{l^3}{EI} = 0.0987654\frac{l^3}{EI}$$

which makes 42% difference.

Galerkin approximation yields also for $\varepsilon = 1$

$$\delta_{12}(1) = \frac{1400}{19685}\frac{l^3}{EI}$$

For another limiting case $\varepsilon \to 0$, we get

$$\delta_{11}(0) = \frac{110,201}{401,241}\frac{l^3}{EI} = 0.2746503\frac{l^3}{EI}$$

instead of exact value $\delta_{11} = (1/3)l^3/EI$.

For $\varepsilon = 0.1$ we get

$$\delta_{11}(0.1) = \frac{110,220}{428,841}\frac{l^3}{EI} = 0.2336996\frac{l^3}{EI}$$

$$\delta_{12}(0.1) = \frac{86,000}{428,841}\frac{l^3}{EI} = 0.200505\frac{l^3}{EI}$$

The natural frequencies are found from Equation (1) by substituting

$y_j = c_j e^{-i\omega t}$ and demanding $c_1^2 + c_2^2 \neq 0$, which yields

$$\begin{vmatrix} 1 - \delta_{11}m\omega^2 & -\delta_{12}m\omega^2 \\ -\delta_{12}m\omega^2 & 1 - \delta_{22}m\omega^2 \end{vmatrix} = 0$$

since $\delta_{12} = \delta_{21}$. Hence, by dividing by $m\omega^2$each of the columns we get

$$\begin{vmatrix} \dfrac{1}{m\omega^2} - \delta_{11} & -\delta_{12} \\ -\delta_{12} & \dfrac{1}{m\omega^2} - \delta_{22} \end{vmatrix} = 0$$

or

$$\left(\frac{1}{m\omega^2} - \delta_{11}\right)\left(\frac{1}{m\omega^2} - \delta_{22}\right) - \delta_{12}^2 = 0 \tag{2}$$

$$\frac{1}{m^2\omega^4} - \frac{1}{m\omega^2}(\delta_{11} + \delta_{22}) + \delta_{11}\delta_{22} - \delta_{12}^2 = 0$$

which results in

$$\frac{1}{m\omega^2} = \frac{\delta_{11} + \delta_{22} \pm \sqrt{(\delta_{11} + \delta_{22})^2 - 4(\delta_{11}\delta_{22} - \delta_{12}^2}}{2}$$

$$= \frac{\delta_{11} + \delta_{22} \pm \sqrt{(\delta_{11} - \delta_{22})^2 + 4\delta_{12}^2}}{2}$$

The natural frequencies are

$$\omega_{I,II}^2 = \frac{2}{m[\delta_{11} + \delta_{22} \pm \sqrt{(\delta_{11} - \delta_{22})^2 + 4\delta_{12}^2}]}$$

For the first mode we have

$$y_1^{(1)} = \lambda_1; \quad \left(\frac{1}{m\omega_I^2} - \delta_{11}\right)\lambda_1 - \delta_{12} y_2^{(1)} = 0$$

$$y_2^{(1)} = \frac{1/m\omega_I^2 - \delta_{11}}{\delta_{12}}\lambda_1$$

For the second mode

$$y_1^{(2)} = \lambda_2; \quad y_2^{(2)} = \frac{1/m\omega_{II}^2 - \delta_{11}}{\delta_{12}}\lambda_2$$

So that the normalized modal matrix is

$$[v] = \begin{bmatrix} \dfrac{\lambda_1}{\mu_1} & \dfrac{\lambda_2}{\mu_2} \\ \dfrac{1/m\omega_I^2 - \delta_{11}}{\delta_{12}} \dfrac{\lambda_1}{\mu_1} & \dfrac{1/m\omega_I^2 - \delta_{11}}{\delta_{12}} \dfrac{\lambda_2}{\mu_2} \end{bmatrix}$$

From Equation (2) also follows that

$$\frac{1/m\omega_I^2 - \delta_{11}}{\delta_{12}} = -\frac{1/m\omega_{II}^2 - \delta_{11}}{\delta_{12}}$$

and $[v]$ can be put in the form

$$[v] = \begin{bmatrix} \dfrac{\lambda_1}{\mu_1} & \dfrac{\lambda_2}{\mu_2} \\ \dfrac{1/m\omega_I^2 - \delta_{11}}{\delta_{12}} \dfrac{\lambda_1}{\mu_1} & -\dfrac{1/m\omega_I^2 - \delta_{11}}{\delta_{12}} \dfrac{\lambda_2}{\mu_2} \end{bmatrix}$$

After normalization $[v]^T[m][v] = I$, one proceeds as in Problem 9.5. For $\varepsilon \to 0$ the results of Problem 9.2 are obtained, with large error if one neglects the cross-correlations.

Exact solution

$$\delta_{11} = \frac{l^3}{3EI}\left[1 - \frac{28}{26+\frac{1}{\varepsilon}} - \frac{3}{8}\left(\frac{2\varepsilon}{2\varepsilon+1} - \frac{42}{26+\frac{1}{\varepsilon}}\right)\right] = \frac{l^3}{3EI}\frac{1+15\varepsilon+8\varepsilon^2}{1+28\varepsilon+52\varepsilon^2}$$

$$\delta_{22} = \delta_{11}$$

$$\delta_{21} = \delta_{12} = \frac{l^3}{6EI}\frac{23\varepsilon+10\varepsilon^2}{1+28\varepsilon+52\varepsilon^2}.$$

Now, since $\delta_{11} = \delta_{22}$ we get

$$\omega_{I,II}^2 = \frac{1}{m(\delta_{11} \pm \delta_{12})}$$

and the analysis proceeds as above.

Derivation of S_{jk}. The system is twice statically indeterminate. We use a standard method of finding R_B and M_B

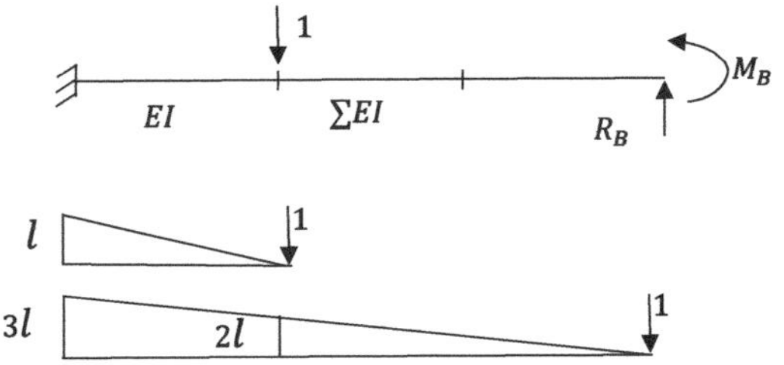

$$\delta_{1B} + R_B\delta_{R_B} + M_B\delta_{M_B} = 0$$

Now $\Delta_B = 0$ & $\theta_B = 0$ or

$$\theta_{1B} + R_B\theta_{R_B} + M_B\theta_{M_B} = 0$$

$$\delta_{1B} = \frac{l^2}{2}\frac{1}{EI}\left(2l + \frac{2}{3}l\right) = \frac{4}{3}\frac{l^2}{EI}$$

Analogically, $\delta_{R_B} = -\frac{l^3}{3EI}\left(20+\frac{7}{\varepsilon}\right)$, $\delta_{M_B} = -\frac{3l^2}{EI}\left(1+\frac{1}{2\varepsilon}\right)$

Finally, we get following equations for R_B & M_B

$$4-\left(20+\frac{7}{\varepsilon}\right)R_B-9\left(1+\frac{1}{2\varepsilon}\right)\frac{M_B}{l}=0$$

$$1-6\left(1+\frac{1}{2\varepsilon}\right)R_B-4\left(1+\frac{1}{2\varepsilon}\right)\frac{M_B}{l}=0$$

yielding $R_B=\frac{7}{26+\frac{1}{\varepsilon}}$, $M_B=(\frac{2\varepsilon}{2\varepsilon+1}-\frac{42}{26+\frac{1}{\varepsilon}})\frac{l}{4}$

Now

$$\delta_{11}=\bar{\delta}_{11}+\bar{\delta}_{B1}R_B+\bar{\delta}_{M1}M_B$$

$$\delta_{12}=\bar{\delta}_{12}+\bar{\delta}_{B2}R_B+\bar{\delta}_{M2}M_B$$

resulting in δ_{jk} as given on page 263.

For $\varepsilon=1$,

$$\delta_{11}(1)=\frac{8}{81}\frac{l^3}{EI},\quad \delta_{12}=\frac{11}{162}\frac{l^3}{EI}$$

For $\varepsilon=0$,

$$\delta_{11}(0)=\frac{27}{81}\frac{l^3}{EI}$$

$$\delta_{12}(0)=0$$

For $\varepsilon\to\infty$

$$\delta_{11}(\infty)=\frac{2}{39}\frac{l^3}{EI}$$

$$\delta_{12}(\infty)=\frac{5}{156}\frac{l^3}{EI}$$

PROBLEM 9.7

Find $E(|X_1|^2)$ and $E(|X_2|^2)$ for the system shown in the figure, where $R_F(\tau)=e^{-\alpha^2\tau^2}$. Use the approximate method for determining the mean-square values.

SOLUTION 9.7

The differential equations of motion read

$$m\ddot{x}_1 + c\dot{x}_1 + (\bar{k} + \varepsilon k)x_1 - \varepsilon k x_2 = F_1(t)$$

$$m\ddot{x}_2 + c\dot{x}_2 + \left(\bar{k} + \varepsilon k\right) x_2 - \varepsilon k x_1 = 0$$

The characteristic equation is

$$\begin{vmatrix} \bar{k} + \varepsilon k - m\omega^2 & -\varepsilon k \\ -\varepsilon k & \bar{k} + \varepsilon k - m\omega^2 \end{vmatrix}$$

or

$$(\bar{k} + \varepsilon k - m\omega^2)^2 - \varepsilon^2 k^2 = 0$$

which yields

$$m\omega^2 - \bar{k} - \varepsilon k = \mp \varepsilon k$$

with

$$m\omega_1^2 = \bar{k}$$

$$m\omega_2^2 = \bar{k} + 2\varepsilon k$$

to find $\{y^1\}$ we solve

$$(\bar{k} + \varepsilon k - m\omega_1^2)y_1^{(1)} = \varepsilon k y_{(2)}^{(1)}$$

which yields for $y_1^{(1)} = \lambda_1$

$$\{v_2^{11}\} = \lambda_1 \begin{Bmatrix} 1 \\ 1 \end{Bmatrix}$$

Analogously

$$(\bar{k} + \varepsilon k - m\omega_2^2)y_1^{(2)} = \varepsilon k y_{(2)}^{(2)}$$

yielding

$$\left\{v_2^{(2)}\right\} = \lambda_2 \begin{Bmatrix} 1 \\ -1 \end{Bmatrix}$$

Finally, with

$$[v]^T [M][v] = [I]$$

we get

$$[v] = \frac{1}{\sqrt{2m}}\begin{bmatrix}1 & 0\\ 0 & 0\end{bmatrix}$$

The excitation spectral matrix is

$$[S_F(\omega)] = \frac{\exp(-\omega^2/4\alpha^2)}{2\alpha\sqrt{\pi}}\begin{bmatrix}1 & 0\\ 0 & 0\end{bmatrix}$$

Now

$$[S_\Phi] = [v]^T[S_F(\omega)][v] = \frac{\exp(-\omega^2/4\alpha^2)}{4m\alpha\sqrt{\pi}}\begin{bmatrix}1 & 1\\ 1 & 1\end{bmatrix}$$

and

$$[S(\omega)] = [H][S_\Phi(\omega)][H]^* = \frac{\exp(-\omega^2/4\alpha^2)}{4m\alpha\sqrt{\pi}}\begin{bmatrix}|H_1|^2 & H_1H_2^*\\ H_1^*H_2 & |H_2|^2\end{bmatrix}$$

with

$$S_{X_1}(\omega) = [v_1][S(\omega)][v_1]^T = \frac{\exp(-\omega^2/4\alpha^2)}{4m\alpha\sqrt{\pi}}[|H_1|^2 + |H_2|^2 + H_1H_2^* + H_1^*H_2]$$

Consider the case when $\varepsilon \gg 1$. Then the cross-correlation terms can be neglected. The approximate method of evaluating the integrals described in Example 9.5, pp. 336–337 yields

$$\begin{aligned}
&E(|X_1|^2)\\
&\approx \int_{-\infty}^{\infty}\frac{\exp(-\omega^2/4\alpha^2)}{8m^2\alpha\sqrt{\pi}}[|H_1(\omega)|^2 + |H_2(\omega)|^2]d\omega\\
&\simeq \frac{\exp(-\omega_1^2/4\alpha^2)}{8m^2\alpha\sqrt{\pi}}\int_{-\infty}^{\infty}|H_1(\omega)|^2 d\omega + |H_1(0)|^2\int_{-\infty}^{\infty}\frac{\exp(-\omega^2/4\alpha^2)}{8m^2\alpha\sqrt{\pi}}d\omega\\
&+\frac{\exp(-\omega_2^2/4\alpha^2)}{8m^2\alpha\sqrt{\pi}}\int_{-\infty}^{\infty}|H_1(\omega)|^2 d\omega + |H_1(0)|^2\int_{-\infty}^{\infty}\frac{\exp(-\omega^2/4\alpha^2)}{8m^2\alpha\sqrt{\pi}}d\omega
\end{aligned}$$

$$= \frac{\exp(-\omega_1^2/4\alpha^2)}{8m^2\alpha\sqrt{\pi}}\frac{1}{2\zeta_2\omega_2^3} + \frac{1}{\omega_2^2}$$

$$+\frac{\exp(-\omega_2^2/4\alpha^2)}{8m^2\alpha\sqrt{\pi}}\frac{1}{2\zeta_2\omega_2^3} + \frac{1}{\omega_2^2}$$

Now for $\varepsilon \ll 1$ we can replace

$$H_2(\omega) \approx H_1(\omega)$$

This results in

$$E(|X_1|^2) \simeq 4[\frac{\exp(-\omega^2/4\alpha^2)}{8m^2\alpha\sqrt{\pi}}\frac{4}{2\zeta_1\omega_1^3} + \frac{1}{\omega_1^2}]$$

$$= \frac{\exp(-\omega^2/4\alpha^2)}{4m^2\alpha\sqrt{\pi}} + \frac{4}{\omega_1^2}$$

PROBLEM 9.8

The system shown in the figure is subjected to ideal white-noise excitation.

(a) Verify that the modal matrix is

$$[v] = \frac{1}{2\sqrt{m}}\begin{bmatrix} 1 & \sqrt{2} & 1 \\ \sqrt{2} & 0 & -\sqrt{2} \\ 1 & -\sqrt{2} & 1 \end{bmatrix}$$

where the corresponding natural frequencies are

$$\omega_1^2 = (2-\sqrt{2})\frac{k}{m}, \quad \omega_2^2 = 2\frac{k}{m}, \quad \omega_3^2 = (2+\sqrt{2})\frac{k}{m}$$

(b) Verify that $E(|X_1|^2) \simeq E(|X_3|^2)$ and $E(|\dot{X}_1|^2) \simeq E(|\dot{X}_3|^2)$.

SOLUTION 9.8

The free body diagrams are

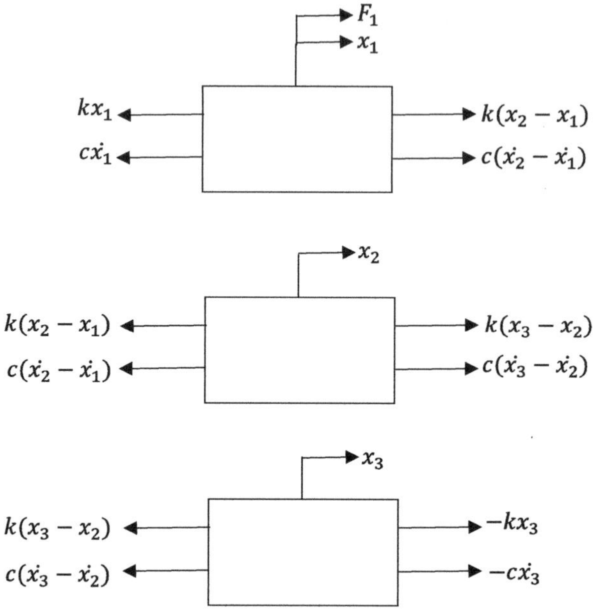

The governing equations are

$$m\ddot{x}_1 + 2c\dot{x}_1 + 2kx_1 - kx_2 - c\dot{x}_2 = F_1(t)$$

$$m\ddot{x}_2 + 2c\dot{x}_2 + 2kx_2 - kx_1 - c\dot{x}_1 - kx_3 - c\dot{x}_3 = 0$$

$$m\ddot{x}_3 + 2c\dot{x}_3 + 2kx_3 - kx_2 - c\dot{x}_2 = 0$$

and

$$[M] = m[I], \quad [K] = k\begin{bmatrix} 2 & -1 & 0 \\ -1 & 2 & -1 \\ 0 & -1 & 2 \end{bmatrix}, \quad [C] = c\begin{bmatrix} 2 & -1 & 0 \\ -1 & 2 & -1 \\ 0 & -1 & 2 \end{bmatrix}$$

The characteristic equation is

$$\begin{vmatrix} 2k - m\omega^2 & -k & 0 \\ -k & 2k - m\omega^2 & -k \\ 0 & -k & 2k - m\omega^2 \end{vmatrix} = 0$$

or

$$[2k - m\omega^2][(2k - m\omega^2)^2 - 2k^2] = 0$$

with

$$\omega_1^2 = (2 - \sqrt{2})k/m$$

$$\omega_2^2 = 2k/m$$

$$\omega_3^2 = \left(2 + \sqrt{2}\right) k/m$$

For the normal modes we can write

$$[K - \omega_j^2 M]\{y^{(j)}\} = \{0\}$$

which yields for $y^{(1)}$

$$\begin{bmatrix} 2k - (2 - \sqrt{2})k & -k & 0 \\ -k & 2k - (2 - \sqrt{2})k & -k \\ 0 & -k & 2k - (2 - \sqrt{2})k \end{bmatrix} \begin{Bmatrix} y_1^{(1)} \\ y_2^{(1)} \\ y_3^{(1)} \end{Bmatrix} = \begin{Bmatrix} 0 \\ 0 \\ 0 \end{Bmatrix}$$

or

$$\{y^{(1)}\} = \lambda_1 \begin{Bmatrix} 1 \\ \sqrt{2} \\ 1 \end{Bmatrix}$$

Accordingly

$$\{y^{(2)}\} = \lambda_2 \begin{Bmatrix} 1 \\ 0 \\ -1 \end{Bmatrix}, \quad \{y^{(3)}\} = \lambda_3 \begin{Bmatrix} 1 \\ -\sqrt{2} \\ 1 \end{Bmatrix}$$

Normalized modal matrix reads

$$[v] = \begin{bmatrix} \dfrac{\lambda_1}{\mu_1} & \dfrac{\lambda_2}{\mu_2} & \dfrac{\lambda_3}{\mu_3} \\ \dfrac{\lambda_1}{\mu_1}\sqrt{2} & 0 & -\dfrac{\lambda_3}{\mu_3}\sqrt{2} \\ \dfrac{\lambda_1}{\mu_1} & -\dfrac{\lambda_2}{\mu_2} & \dfrac{\lambda_3}{\mu_3} \end{bmatrix}$$

requirement

$$[v][m][v]^T = [I]$$

yields

$$\left(\frac{\lambda_1^2}{\mu_1^2}+\frac{\lambda_2^2}{\mu_2^2}+\frac{\lambda_3^3}{\mu_3^3}\right)m=1$$

$$2\left(\frac{\lambda_1}{\mu_1}\right)^2+2\left(\frac{\lambda_3}{\mu_3}\right)^2=1$$

$$\left(\frac{\lambda_1}{\mu_1}\right)^2 2m-\left(\frac{\lambda_2}{\mu_2}\right)^2 2m+\left(\frac{\lambda_3}{\mu_3}\right)^2 2m=0$$

which is in turn results in

$$\frac{\lambda_1}{\mu_1}=\frac{\lambda_3}{\mu_3}=\frac{1}{2\sqrt{m}},\quad \frac{\lambda_2}{\mu_2}=\frac{\sqrt{2}}{2\sqrt{m}}$$

Finally

$$[v]=\frac{1}{2\sqrt{m}}\begin{bmatrix}1 & \sqrt{2} & 1\\ \sqrt{2} & 0 & -\sqrt{2}\\ 1 & -\sqrt{2} & 1\end{bmatrix}=[v]^T$$

Now

$$[S_F(\omega)]=S_0\begin{bmatrix}1 & 0 & 0\\ 0 & 0 & 0\\ 0 & 0 & 0\end{bmatrix}$$

and

$$[S_\Phi(\omega)]=[v]^T[S_F][v]=\frac{S_0}{4m}\begin{bmatrix}1 & \sqrt{2} & 1\\ \sqrt{2} & 2 & \sqrt{2}\\ 1 & \sqrt{2} & 1\end{bmatrix}$$

Let us find (see Equation 9.161)

$$[S_\Phi(\omega)]=[H(\omega)][S_\Phi(\omega)][H(\omega)]^*$$

$$=\frac{S_0}{4m}\begin{bmatrix}|H_1|^2 & \sqrt{2}H_1H_2^* & H_1H_3^*\\ \sqrt{2}H_2H_1^* & 2|H_2|^2 & \sqrt{2}H_2H_3^*\\ H_3H_1^* & \sqrt{2}H_3H_2^* & |H_3|^2\end{bmatrix}$$

Now, in accordance with 9.161

$$S_{X_i} = [v_i][S][v_i]^T$$

resulting in

$$S_{X_1} = \frac{S_0}{16m^2}[(H_1)^2 + 4|H_2|^2 + |H_3|^2 + 2\text{Re}\left(H_1 H_3^*\right) + 4\text{Re}\left(H_1 H_2^* + H_2 H_3^*\right)]$$

$$S_{X_3} = \frac{S_0}{16m^2}[(H_1)^2 + 4|H_2|^2 + |H_3|^2 + 2\text{Re}(H_1 H_3^*)$$

Since

$$\int_{-\infty}^{\infty} |H_j|^2 d\omega = \frac{\pi}{2\zeta_j \omega_j^3}$$

and
$2\int_{-\infty}^{\infty} \text{Re} H_i H_j^* d\omega$ is given by Equation (9.198) we get

$$E(|X_1|^2) = \frac{S_0\pi}{ck}\left[\frac{1}{(2+\sqrt{2})^2} + 1 + \frac{1}{2-\sqrt{2})^2} + \frac{2(c^2/km)}{1+\zeta(c^2/km)} + \frac{64c^2/km}{2+144c^2/km}\right]$$

$$E(|X_3|^2) = \frac{S_0\pi}{ck}\left[\frac{1}{(2+\sqrt{2})^2} + 1 + \frac{1}{2-\sqrt{2})^2} + \frac{2(c^2/km)}{1+\zeta(c^2/km)} - \frac{64c^2/km}{2+144c^2/km}\right]$$

For, say, $c^2/km = 0.01$, we get

$$E(|X_1|^2) = \frac{S_0\pi}{ck}(4.02 + 0.18) = 4.2\frac{S_0\pi}{ck}$$

$$E(|X_3|^2) = \frac{S_0\pi}{ck}(4.02 - 0.18) = 3.84\frac{S_0\pi}{ck}$$

Therefore approximately

$$E(|X_1|^2) \simeq E(|X_2|^2)$$

Results for the mean square velocities are

$$E(|\dot{X}_1|^2) = \frac{S_0\pi}{cm}\left[\frac{1}{2+\sqrt{2}} + 2 + \frac{1}{2-\sqrt{2}} + \frac{16c^2/km}{1+5c^2/km} \pm \frac{134c^2/km}{2+144c^2/km}\right]$$

and finally

$$E(|\dot{X}_1|^2) = \frac{S_0\pi}{cm}(4.15+0.39) = 4.54\frac{S_0\pi}{cm}$$

$$E(|\dot{X}_3|^2) = \frac{S_0\pi}{cm}(4.15-0.39) = 3.76\frac{S_0\pi}{cm}$$

The asymmetry here is more pronounced, and the neglect of the cross-correlation term for $E(|X_1^2|)$ entails an error of 8.6%, whereas for $E(|X_3^2|)$ it constitutes 10.37%.

PROBLEM 9.9

An n-degrees-of-freedom system is subjected to ideal white-noise loading with intensity S_0. Each mass is connected both with the ground and with the other mases, the springs being identical with stiffness k. Damping proportional to the mass is provided by a dashpot attached to each mass, the damping coefficient being c.

(a) Verify that this system possesses $n-1$ identical natural frequencies.
(b) Estimate $E(|X_1|^2)$ in the particular case $n=4$ (for $n=3$ the problem reduces to Example 9.11.)

SOLUTION 9.9

(a) The equations of motion read

$$[M]\{\ddot{x}\} + [C]\{\dot{x}\} + [K]\{x\} = \{F\}$$

where

$$[M] = [m], \quad [C] = [c]$$

and

$$[K] = \begin{bmatrix} nk & -k & -k & \cdots & -k & -k \\ -k & nk & -k & \cdots & -k & -k \\ -k & -k & nk & & -k & -k \\ \vdots & & & & & \\ -k & -k & -k & \cdots & -k & nk \end{bmatrix}$$

The natural frequencies are obtained from

$$\det[K - \omega^2 M] = 0$$

or, in other words from

$$\det\begin{bmatrix} nk - m\omega^2 & -k & \cdots & -k & -k \\ -k & nk - m\omega^2 & \cdots & -k & -k \\ -k & -k & & -k & -k \\ \vdots & \vdots & & & \\ -k & -k & \cdots & -k & nk - m\omega^2 \end{bmatrix} = 0$$

Substituting the last column successively from the other columns we obtain

$$\det\begin{bmatrix} (n+1)k - m\omega^2 & 0 & \cdots & 0 & -k \\ 0 & -(n+1)k + m\omega^2 & \cdots & 0 & -k \\ 0 & 0 & & 0 & -k \\ \vdots & \vdots & & & \\ -(n+1)k & -(n+1)k & & -(n+1)k & nk \\ +m\omega^2 & -m\omega^2 & & +m\omega^2 & +m\omega^2 \end{bmatrix} = 0$$

The all rows are now added to the last one to yield

$$\det\begin{bmatrix} (n+1)k - M\omega^2 & 0 & \cdots & 0 & -k \\ 0 & (n+1)k - M\omega^2 & \cdots & 0 & -k \\ 0 & 0 & & 0 & -k \\ \vdots & \vdots & & \vdots & \\ 0 & 0 & & 0 & k - m\omega^2 \end{bmatrix} = 0$$

Expanding the above determinant with respect to the last row, we obtain

$$[(n+1)k - m\omega^2]^{n-1}(k - m\omega^2) = 0$$

The natural frequencies are thus represented by two sets of eigenfrequencies: one comprising the separate first natural frequency

$$\omega_1^2 = k/m$$

and the other remaining identical $n-1$ natural frequencies

$$\omega_2^2 = \omega_3^2 = \cdots = \omega_n^2 = \frac{(n+1)k}{m}$$

For $n=4$, we get

$$[V] = \frac{1}{\sqrt{m}} \begin{bmatrix} \frac{1}{2} & \frac{1}{\sqrt{2}} & \frac{1}{\sqrt{6}} & \frac{1}{\sqrt{12}} \\ \frac{1}{2} & -\frac{1}{\sqrt{2}} & \frac{1}{\sqrt{6}} & \frac{1}{\sqrt{12}} \\ \frac{1}{2} & 0 & -\frac{2}{\sqrt{6}} & \frac{1}{\sqrt{12}} \\ \frac{1}{2} & 0 & 0 & -\frac{3}{\sqrt{12}} \end{bmatrix}$$

Substitution

$$\{x\} = [V]\{q\}$$

yields

$$\{\ddot{q}\} + \frac{c}{m}\{\dot{q}\} + [\omega^2]\{q\} = [V]^T\{F\}$$

Further analysis (for any n) can be found in the paper entitled "Random Vibration of System with Finitely Many Degrees of Freedom and Several Coalescent Natural Frequencies," by E. Lubliner and I. Elishakoff, Int. Journal of Engineering Science, 1986, Vol. 24(4), pp. 461–470.

chapter 10

Random Vibration of Continuous Structures

PROBLEM 10.1

An infinite plate is subjected to random loading represented by a space-wise fully correlated random function, $R_q(\tau, \xi_1, \xi_2) = R_q(\tau)$. Find the mean-square displacement and acceleration.

SOLUTION 10.1

For fully correlated random function, Eq. (10.13) yields

$$s_q(w, k_1, k_2) = \frac{1}{(2\pi)^3} \iiint_{-\infty}^{\infty} R_q(\tau, \xi_1, \xi_2)$$

$$\exp\left[-i\left(\omega\tau + \kappa_1\xi_1 + \kappa_2\xi_2\right)\right] d\tau d\xi_1 d\xi_2$$

$$= \frac{1}{2\pi} \int_{-\infty}^{\infty} R_q(\tau)\, e^{-i\omega\tau} d\tau \cdot$$

$$\frac{1}{(2\pi)^2} \iint_{-\infty}^{\infty} e^{-i(\kappa_1\xi_1 + \kappa_2\xi_2)} d\xi_1 d\xi_2$$

$$= S_q(\omega)\delta(\kappa_1)\delta(\kappa_2)$$

Equation (10.20) yields then

$$\begin{aligned} R_w(0) &= R_w(0,0,0) \\ &= \iiint_{-\infty}^{\infty} \frac{1}{\left|D\left(\kappa_1^2+\kappa_2^2\right)^2 - \rho h\omega^2\right|^2} S_q(\omega,\kappa_1,\kappa_2)\, d\omega d\kappa_1 d\kappa_2 \\ &= \iiint_{-\infty}^{\infty} \frac{1}{\left|D\left(\kappa_1^2+\kappa_2^2\right)^2 - \rho h\omega^2\right|^2} S_q(\omega)\,\delta(\kappa_1)\,\delta(\kappa_2)\, d\omega d\kappa_1 d\kappa_2 \end{aligned}$$

For the mean-square acceleration the integrand of this expression is multiplied by ω^4.

PROBLEM 10.2

An infinite plate is subjected to random loading, represented by space-wise white noise and time-cut white noise. Find the mean-square acceleration.

SOLUTION 10.2

Now

$$R_q(\tau,\xi_1,\xi_2) = R_q(\tau)\delta(\xi_1)\delta(\xi_2)$$

but

$$\begin{aligned} S_q(\omega) &= \frac{1}{2\pi}\int_{-\infty}^{\infty} R_q(\tau)e^{-i\omega\tau}\,dt \\ &= \begin{cases} S_0, & \text{for } |\omega| \le \omega_c \\ 0, & \text{otherwise} \end{cases} \end{aligned}$$

Instead of Equation 10.24 we get

$$R_w(0) = \int_{-\omega_c}^{\omega_c} \frac{\pi^2 S_q(\omega)\,d\omega}{2\rho h|\omega^3\mu|(D_r\rho h)^{1/2}}$$

and for the mean-square acceleration the integrand is multiplied by ω^4.

PROBLEM 10.3

Apply the normal mode method for a beam under a distributed random load, simply supported at one end and clamped at the other.

SOLUTION 10.3

We attach the origin of the coordinates with the simply supported end ($x = 0$).
Instead of Equation 10.34, we get

$$\begin{vmatrix} 0 & 1 & 0 & 1 \\ 0 & -r^2 & 0 & r^2 \\ S & C & S & C \\ rc & -rs & rC & rS \end{vmatrix} = 0$$

i.e. we compose our determinant from the first two rows from determinant (10.35) and the two last ones from the last equation on p. 391. The characteristic equation is

$$sC - cS = 0$$

or

$$\tan(rl) = \tanh(rl)$$

with asymptotic roots

$$r_j^l \approx \left(j + \frac{1}{4}\right)$$

Further analysis goes as in 10.2.1 and 10.2.4.

PROBLEM 10.4

Put Eq. (10.53) in the form

$$w\,(x,t) = \int_{-\infty}^{\infty} d\tau \int_0^l q\,(x, t-\tau)\,h\,(x,\tau,\xi)\,d\xi$$

where

$$h\,(x,\tau,\xi) = \frac{1}{\rho A}\sum_{j=1}^{\infty} \psi_j\,(x)\,\psi_j\,(\xi)\,h_j(\tau)\upsilon_j^{-2}$$

and show that $h(x, \tau, \xi)$ is the response to excitation in the form $\delta(x - \xi)\delta(t - \tau)$. [$h(x, \tau, \xi)$ is called the*unit impulse response function* or *Green function* of the system.]

Note to lecturer: Equation for $w(x, t)$, change the argument of q by ξ instead of x.

SOLUTION 10.4

Indeed, upon substitution

$$h(x, \tau, \xi) = \frac{1}{\rho A} \sum_{j=1}^{\infty} \psi_j(x)\psi_j(\xi)h_j(\tau)\upsilon_j^{-2}$$

we get

$$w(x, t) = \int_{-\infty}^{\infty} d\tau \int_0^l q(x, t - \tau)\, h(x, \tau, \xi)\, d\xi \tag{1}$$

Now, let us consider a standard excitation of the impulse applied at the position $x = \xi$ at the time instant τ, i.e.

$$q(x, t) = \delta(x - \xi)\delta(t - \tau) \tag{2}$$

with zero initial conditions. Denote the displacement response of this excitation by

$$w(x, t) = h(x, t - \tau, \xi)$$

here $h(x, t, \xi)$ is the impulse response function. Then the response to the arbitrary excitation $q(x, t)$ is easily obtained by the superposition, applicable to the linear system. Since (3) is response to (2), then the response to arbitrary excitation is

$$w(x, t) = \int_{-\infty}^{\infty} d\tau \int_0^l q(\xi, t - \tau)\, h(x, \tau, \xi)\, d\xi$$

as the response to

$$q(x, t) = \int_{-\infty}^{\infty} d\tau \int_0^l q(\xi, t - \tau)\, \delta(x - \xi)\, \delta(\tau)\, d\xi, \quad \text{Q.E.D.}$$

PROBLEM 10.5

A cantilever beam is subjected to random loading with autocorrelation function as in Eq. (10.80). Find a closed expression for the mean-square value of the displacement at the free end.

Note to lecturer: The problem should read, "A cantilever is subjected to random loading with auto correlation function as in 10.86."

SOLUTION 10.5

Under these boundary conditions we have in Equation 10.29

$$P_1 = 1 \quad P_3 = \frac{\partial^2}{\partial x^2}$$

$$P_2 = \frac{\partial}{\partial x} \quad P_4 = \frac{\partial^3}{\partial x^3}$$

so that instead of (10.35) we obtain

$$\begin{vmatrix} 0 & 1 & 0 & 1 \\ r & 0 & r & 0 \\ -r^2 s & -r^2 c & r^2 C & r^2 C \\ -r^3 c & -r^3 s & r^3 C & r^3 S \end{vmatrix} = 0$$

which yields in

$$1 + cC = 0$$

with asymptotic solutions

$$r_j^l = \frac{1}{2}(2j - 1)\pi$$

and

$$\psi_j(x) = \sin r_j x - \sin h r_j x + A_j(\cos h r_j x - \cos r_j x)$$

with

$$A_j = \frac{\sin r_j l + \sin h r_j l}{\cos r_j l + \cos h r_j l}$$

Now in accordance with Lord Rayleigh, "Theory of Sound", Vol. 2, p. 365, we have

$$v_j^2 = \int_0^l \psi_j^2 (x) \, dx = \frac{l}{4} [\psi_j (l)]^2$$

Simple calculation shows that

$$\psi_j (l) = 1$$

Hence

$$v_j^2 = \frac{l}{4}$$

The cross-spectral densities are, instead of Equation (10.87)

$$S_{Q_j Q_k} = \frac{1}{v_j^2 v_k^2} \int_0^l \int_0^l \frac{R}{2\pi l} \delta(x_2 - x_1)\, \psi_j(x_1)\, \psi_k(x_2)\, dx_1 dx_2$$

$$= \frac{16}{l^2} \frac{R}{2\pi l} \int_0^l \psi_j(x_1)\, \psi_k(x_1)\, dx_1$$

$$= \frac{16}{l^2} \cdot \frac{R}{2\pi l} \cdot \frac{l}{4} \delta_{jk} = \frac{2R}{\pi l^2} \delta_{jk}$$

and

$$R_w(x, x, 0) = \frac{2R}{\pi(\rho A l)} \sum_{j=1}^{\infty} \psi_j^2(x) \int_{-\infty}^{\infty} \frac{d\omega}{\left(\omega_j^2 - \omega^2\right)^2 + 4\zeta_j \omega_j^2 \omega^2}$$

$$= \frac{2R}{\pi(\rho A l)^2} \cdot \frac{\pi \rho A}{c} \sum_{j=1}^{\infty} \frac{1}{\omega_j^2} \psi_j^2(x)$$

for the free end

$$R_w(l, l, 0) = \frac{2R}{c\rho A l} \sum_{j=1}^{\infty} \frac{1}{\omega_j^2} \psi_j^2(l)$$

but,

$$\psi_j^2(l) = 1$$

and

$$R_w(l, l, 0) = \frac{2R}{c\rho A l} \sum_{j=1}^{\infty} \frac{1}{\omega_j^2}$$

However, (see p. 390)

$$\omega_j^2 = r_j^4 \frac{EI}{PA} = (r_j l)^4 \frac{EI}{\rho A l^4}$$

and

$$R_w(l, l, 0) = \frac{2R}{c\rho A l} \frac{\rho A l^4}{EI} \sum_{j=1}^{\infty} \frac{1}{(r_j l)^4}$$

$$= \frac{2R l^3}{cEI} \sum_{j=1}^{\infty} \frac{1}{(r_j l)^4}.$$

PROBLEM 10.6

Derive expression (10.99).

SOLUTION 10.6

For the mean- square displacement we obtain Eq. (10.97)

$$E[w^2(x,t)] = \sum_{j=1}^{\infty}\sum_{k=1}^{\infty} \psi_j(x)\psi_k(\alpha)\psi_j(a)\psi_k(a)x\nu_j^{-2}\nu_k^{-2}I_{jk}$$

where

$$I_{jk} = \int_{-\infty}^{\infty} S_F(\omega)H_j^*(\omega)H_k(\omega)d\omega$$

For the mean-square velocity $E[v^2(x,t)]$ the derivation are completely analogous. Instead of the expansion on page 399 of the textbook, the equation below Eq. (10.68) reads for the displacement

$$w(x,t) = \sum_{j=1}^{\infty} \psi_j(x) \int_{-\infty}^{\infty} W_j(\omega)e^{i\omega t}d\omega$$

For the velocity

$$v(x,t) = \frac{dw}{dt}$$

we use analogous expansion

$$v(x,t) = \sum_{j=1}^{\infty} \psi_j(x) \int_{-\infty}^{\infty} V_j(\omega)e^{iwt}$$

where

$$V_j(\omega) = i\omega W_j(\omega)$$

Therefore, for the mean-square velocity we obtain expression

$$E[v^2(x,t)] = \sum_{j=1}^{\infty}\sum_{k=1}^{\infty} \psi_j(x)\psi_k(x)\psi_j(x)\psi_k(a)\nu_j^{-2}\nu_k^{-2}I'_{jk}$$

where

$$I'_{jk} = \int_{-\infty}^{\infty} \omega^2 S_f(\omega)H_j^*(\omega)H_k(\omega)d\omega$$

as the problem stipulates.

PROBLEM 10.7

Show that the span average of $G(x/l,\ a/l,\ N_c)$ in Sec. 10.4 is $g(a/l,\ N_c)$, and the average with respect to the loading positions of the span average is N_c. That is,

$$\frac{1}{l^2}\int_0^l dx \int_0^l G\left(\frac{x}{l},\ \frac{a}{l},\ N_c\right) da = N_c$$

SOLUTION 10.7

In accordance with Equation 10.109

$$G\left(\frac{x}{3},\ \frac{a}{l},\ N_c\right) = g\left(\frac{x}{l},\ N_c\right) + g(\frac{a}{l},\ N_c)$$

$$-\frac{1}{2}g\left(\frac{x-a}{l},\ N_c\right) - \frac{1}{2}g(\frac{x-l+a}{l},\ N_c)$$

we should show that

$$\frac{1}{l}\int_0^l G\left(\frac{x}{l},\ \frac{a}{l},\ N_c\right) dx = g\left(\frac{a}{l},\ N_c\right)$$

since

$$\frac{1}{l}\int_0^l g\left(\frac{a}{l},\ N_c\right) dx = g\left(\frac{a}{l},\ N_c\right)$$

we should show that the span average of the first, the third and the fourth terms is zero:

$$\frac{1}{l}\int_0^l \left[N_c + \frac{1}{2} - \frac{1}{2}\frac{\sin\,(2N_c+1)\,\pi x}{\sin\pi x}\right] dx$$

$$-\frac{1}{2l}\int_0^l \left[N_c + \frac{1}{2} - \frac{1}{2}\frac{\sin(2N_c+1)\pi\,(x-a)/l}{\sin\pi\,(x-a)/l}\right] dx$$

$$-\frac{1}{2l}\int_0^l \left[N_c + \frac{1}{2} - \frac{1}{2}\frac{\sin(2N_c+1)\pi\,(x+a-l)/l}{\sin\pi\,(x+a-l)/l}\right] dx \underline{?}0$$

We denote

$$\frac{x-a}{l} = z$$
$$\frac{x+a-l}{l} = y$$

and

$$\frac{1}{l}\int_0^l \left[N_c + \frac{1}{2} - \frac{1}{2}\frac{\sin\left(2N_c+1\right)\pi x}{\sin \pi x}\right]dx$$
$$-\frac{1}{2}\int_{-a/l}^{1-a/l}\left[N_c + \frac{1}{2} - \frac{1}{2}\frac{\sin\left(2N_c+1\right)\pi z}{\sin \pi z}\right]dz$$
$$-\frac{1}{2}\int_{a/l-1}^{a/l}\left[N_c + \frac{1}{2} - \frac{1}{2}\frac{\sin\left(2N_c+1\right)\pi y}{\sin \pi y}\right]dy$$

According to Gradshteyn, I.S. and Ryzhik, I.M., "Table of Integrals, Series and Products", Academic Press, New York, 1980, Equation 2/539.1. We have

$$\int \frac{\sin\left(2N+1\right)x}{\sin x}dx = 2\sum_{k=1}^{N}\frac{\sin 2kx}{2k} + x$$

and what is left is

$$\left(N_c + \frac{1}{2} - \frac{1}{2}\right) - \frac{1}{2}\left[\left(N_c + \frac{1}{2}\right)\left(1 - \frac{a}{l} + \frac{a}{l}\right)\right]$$
$$-\frac{1}{2}\left[\left(N_c + \frac{1}{2}\right)\left(\frac{a}{l} + 1 - \frac{a}{l}\right)\right] + \frac{1}{2}\cdot\frac{1}{2}\left(1 - \frac{a}{l} + \frac{a}{l}\right)$$
$$+\frac{1}{2}\cdot\frac{1}{2}\left(\frac{a}{l} + 1 - \frac{a}{l}\right) \equiv 0$$

Now

$$\frac{1}{l}\int_0^l g\left(\frac{a}{l}, N_c\right)da = \frac{1}{l}\int_0^l \left[N_c + \frac{1}{2} - \frac{1}{2}\frac{\sin\left(2N_c+1\right)\pi a/l}{\sin \pi a/l}\right]da$$
$$= N_c + \frac{1}{2} - \frac{1}{2} = N_c$$

therefore,

$$\frac{1}{l^2}\int_0^l dx \int_0^l G\left(\frac{x}{l}, \frac{a}{l}, N_c\right) dx$$

$$= \frac{1}{l}\int_0^l g\left(\frac{a}{l}, N_c\right) da = N_c, \text{ Q.E.D.}$$

PROBLEM 10.8

Use the normal mode approach to derive the mean-square values of the displacement, velocity, and acceleration of a beam of span l, clamped at both ends and subjected to a concentrated load $F(t)$ at the cross section $x = a$. $F(t)$ is band-limited white noise with ω_c as the cutoff frequency. Examine the behavior of the responses as ω_c increases and tends to infinity.

SOLUTION 10.8

According to Equation (10.96)

$$S_{Q_jQ_k}(\omega) = \psi_j(a)\psi_k(a)\frac{1}{\upsilon_j^2\upsilon_k^2}S_F(\omega)$$

where

$$\upsilon_j^2 = \int_0^l \psi_j^2(x)\,dx$$

In accordance with Equation (10.40)

$$\psi_j(x) = \sin r_j x - \sinh r_j x + A_j(\cos r_j x - \cosh r_j x)$$

$$A_j = \frac{\cos r_j l - \cosh r_j l}{\sin r_j l + \sinh r_j l}$$

$$r_j(l) \simeq \left(j + \frac{1}{2}\right)\pi$$

As is shown by Lord Rayleigh, "Theory of Sound", Dover, New York, p. 265,

$$\int_0^l \psi_j^2(x)\,dx = \frac{l}{4}[\psi_j''(l)]$$

Now

$$\psi_j''(l) = r_j^2[-\sin r_j l - \sinh r_j l + A_j(-\cos r_j l - \cosh r_j l)]$$

$$A_j \simeq -1$$

$$v_j^2 = \frac{l}{4}\psi_j''(l)$$

Hence

$$S_{Q_jQ_k} = \frac{\psi_j(a)\,\psi_k(a)}{\psi_j''(l)\,\psi_k''(l)}\frac{4}{l^2}S_F(\omega)$$

Further steps are as in Section 10.5, Equation (10.97), with simplification following from Equation (10.39)

$$\omega_j^2 \simeq \frac{EI}{\rho A}\left(j + \frac{1}{2}\right)^4 \frac{\pi^4}{l^4}.$$

PROBLEM 10.9

A nonuniform beam is simply supported at both ends and subjected to a concentrated force at the cross section $x = a$, $a < l$. $F(t)$ is a band-limited white noise with ω_c as the cutoff frequency. Find the mean-square velocity as a function of the parameter α. Evaluate the span average mean-square velocity as a function of α (see figure).

SOLUTION 10.9

The differential Equation (10.25) takes form

$$EI\left[1 + (\alpha - 1) < x - l >^{\circ}\right]\frac{\partial^4 w}{\partial x^4} + c\frac{\partial w}{\partial t} + \rho A\frac{\partial^2 w}{\partial t^2} = q(x, t)$$

Let us assume that ω_c is slightly above the first natural frequency; then only the first mode contribution will be of significance (see Section 10.5). Therefore one-term Galerkin method (see for its description appendix E, pp. 477–482) will be justified; we assume therefore

$$w(x, t) = Y_1(t)\sin\frac{\pi x}{2l} = Y_1(t)\psi_1(x)$$

substitution in the governing equation yields an "error"

$$\varepsilon(x,t) = EI[1+(\alpha-1)\langle x-l\rangle^{\circ}]\left(\frac{\pi}{2l}\right)^4 \sin\left(\frac{\pi x}{2l}\right) Y_1(t) + c\sin\left(\frac{\pi x}{2l}\right)\dot{Y}_1(t) + \rho A \sin\left(\frac{\pi x}{2l}\right)\ddot{Y}_1(t) - q(x,t)$$

We require orthogonality $(\varepsilon, \psi_1) = 0$, yielding

$$EI\left(\frac{\pi}{2l}\right)^4 lY_1(t) + EI(\alpha-1)\left(\frac{\pi}{2l}\right)^4 \int_l^{2l} \sin^2\left(\frac{\pi x}{2l}\right) dx + cl\dot{Y}_1 + \rho Al\ddot{Y}_1(t) = \int_0^{2l} q(x,t)\sin\frac{\pi x}{2l} dx$$

$$\ddot{Y}_1 + \frac{c}{\rho A}\dot{Y}_1 + \frac{EI}{\rho A}\left(\frac{\pi}{2l}\right)^4 Y_1 + \frac{EI(\alpha-1)}{\rho A}\left(\frac{\pi}{2l}\right)^4 \frac{1}{l}\int_l^{2l} \sin^2\left(\frac{\pi x}{2l}\right) dx$$

$$\frac{1}{\rho Al}\int_0^{2l} q(x,t)\sin\left(\frac{\pi x}{2l}\right) dx \tag{1}$$

for free vibration frequency, we put $c \equiv 0,\ q(x,t) \equiv 0$, to yield after substitution

$$Y(t) = e^{-i\omega_1 t}$$

$$\omega_1^2 = \frac{EI}{\rho A}\left(\frac{\pi}{2l}\right)^4\left[1+(\alpha-1)\frac{1}{l}\int_l^{2l}\sin^2\left(\frac{\pi x}{2l}\right)dx\right] = \frac{EI}{\rho A}\left(\frac{\pi}{2l}\right)^4 \left[1+\frac{\alpha-1}{2}\right] = \frac{\alpha+1}{2}\frac{EI}{\rho A}\left(\frac{\pi}{2l}\right)^4 \equiv \frac{\alpha+1}{2}\omega_1^{-2} \tag{2}$$

were $\bar{\omega}_1$ is the natural frequency of homogeneous beam with stiffness EI. In view of Equation (2), Equation (1) becomes

$$\ddot{Y}_1 + \frac{c}{\rho A}\dot{Y}_1 + \omega_1^2 Y_1 = \frac{1}{\rho Al}\int_0^{2l} q(x,t)\sin\left(\frac{\pi x}{2l}\right) dx$$

$$q(x,t) = F(t)\delta(x-a)$$

and

$$\ddot{Y}_1 + \frac{c}{\rho A}\dot{Y}_1 + \omega_1^2 Y_1 = \frac{1}{\rho Al}\sin\left(\frac{\pi a}{2l}\right) F(t)$$

Since $\omega_c > \omega_0$, one can assume $F(t)$ to be ideal white noise, with $S_F \equiv S_0$. With notation (10.51)

$$\zeta_1 = \frac{c}{2\rho A \omega_1}$$

we have

$$\ddot{Y}_1 + 2\zeta_1\omega_1\dot{Y}_1 + \omega_1^2 Y_1 = \frac{1}{l}\sin\left(\frac{\pi A}{2l}\right)F(t)$$

and the mean-square displacement becomes (see Equation 9.67)

$$E\left[Y_1^2(t)\right] = \frac{S_0\pi}{2\zeta\omega_1^3}\frac{1}{(\rho A l)^2}\sin^2\left(\frac{\pi a}{2l}\right)$$

and

$$\begin{aligned}
E\left[w^2(x,t)\right] &= E\left[Y_1^2(t)\right]\sin^2\left(\frac{\pi x}{2l}\right) \\
&= \frac{S_0\pi}{\omega_1^2 c\rho A l^2}\sin^2\left(\frac{\pi a}{2l}\right)\sin^2\left(\frac{\pi x}{2l}\right) \\
&= \frac{2}{\alpha+1}E[\bar{w}^2(x,t)]
\end{aligned}$$

where $E[\bar{w}^2(x,t)]$ is a mean-square response of the homogeneous beam with stiffness EI.

Span-average response is

$$\begin{aligned}
S &= \frac{1}{2l}\int_0^{2l} E\left[w^2(x,t)\right]dx = \frac{S_0\pi}{\omega_1^2 c\rho A l}\sin^2\left(\frac{\pi a}{2l}\right) \\
&= \frac{2}{\alpha+1}\bar{S}
\end{aligned}$$

where $\bar{S}$ is a span-average response of the homogeneous beam with stiffness EI; with $\alpha \to \infty$, $s \to 0$.

PROBLEM 10.10

Generalize the results of Sec. 10.4 to a plate on an elastic foundation (see figure).

SOLUTION 10.10

The differential equation reads

$$D\left(\frac{\partial^4 w}{\partial x^4}+2\frac{\partial^4 w}{\partial x^2\partial y^2}+\frac{\partial^4 w}{\partial y^4}\right)+c\frac{\partial w}{\partial t}+\rho h\frac{\partial^2 w}{\partial t^2}+kw=q(x,q,t) \quad (1)$$

where

$$D=\frac{Eh^3}{12(1-\upsilon^2)}$$

is cylindrical stiffness, E is Young's modulus, his thickness, υ-Possion's ratio, x, y_1 are the coordinates.

For free vibrations, we have

$$D\left(\frac{\partial^4 w}{\partial x^4}+2\frac{\partial^4 w}{\partial x^2\partial y^2}+\frac{\partial^4 w}{\partial y^4}\right)+k\omega+\rho h\frac{\partial^2 w}{\partial t^2}=0 \quad (2)$$

For simply supported plate the boundary conditions read

$$x=0,\ x=a:\ w=\frac{\partial^2 w}{\partial x^2}=0$$

$$y=0,\ y=b:\ w=\frac{\partial^2 w}{\partial y^2}=0$$

These conditions are satisfied if we put

$$w\,(x,y,t)=A\sin\frac{m\pi x}{a}\sin\frac{n\pi y}{b}e^{i\omega t} \quad (3)$$

where m is the number of half waves in x-direction, n is the number of half waves in y direction "a" is the length of the side in x direction, and "b" is the length of the side in y direction. Substitution of (3) into (2) yields

$$A\left[D\left(\frac{m^2\pi^2}{a^2}+\frac{n^2\pi^2}{b^2}\right)^2+k-\rho h\omega^2\right]=0$$

Due to nontrivality of A, we get expression for the natural frequency squared

$$\omega_{mn}^2=\frac{D}{\rho A}\left(\frac{m^2+\pi^2}{a^2}+\frac{n^2\pi^2}{b^2}\right)^2+k \quad (4)$$

which can assume only certain values given by (4), with m and n — integers. Therefore we denote ω by ω_{mn}:

$$\omega_{mn}^2=\frac{D}{\rho A}\left(\frac{m^2\pi^2}{a^2}+\frac{n^2\pi^2}{b^2}\right)=\frac{D\pi^4}{\rho A d^4}\left(m^2+n^2\frac{a^2}{b^2}\right)^2+k$$

For plates with a/b = integer we might get coinciding natural frequencies. Consider case of the square plate, $a = b$, then the nondimensional natural frequency Ω_{mn}

$$\Omega_{mn}^2 = \omega_{mn}^2 \frac{\rho A a^4}{D\pi^4}$$

becomes

$$\Omega_{mn}^2 = (m^2 + n^2)^2 + \frac{k\rho A a^4}{D\pi^4}$$

We introduce the nondimensional elastic foundation parameter

$$K = \frac{k\rho A a^4}{D\pi^4}$$

Then

$$\Omega_{mn}^2 = (m^2 + n^2)^2 + K$$

consider first case of foundationless plate. Then

$$\Omega_{mn}^2 = (m^2 + n^2)^2$$

we have double frequencies of type

$$\Omega_{mn} = \Omega_{nm}$$

In particular $\Omega_{12} = \Omega_{21}$, $\Omega_{13} = \Omega_{31}$, $\Omega_{14} = \Omega_{41}$ etc. In addition we may have triple frequencies

$$\Omega_{17} = \Omega_{71} = \Omega_{55} = 50^2 = 2,500$$

quadruple frequencies and frequencies of higher multiplicity. The same happens also for the plate on elastic foundation. In the latter case, in addition, if K is sufficiently large,

$$K \gg \left(1^2 + 1^2\right)^2 = 4$$

one may get very close natural frequencies too. Say, $K = 10$. Then

$$\Omega_{11} = 1004,\ \Omega_{12} = \Omega_{21} = 1025,\ \Omega_{13} = \Omega_{31} = 1100, \text{etc.}$$

The normal mode analysis, applied to plates should not neglect cross-correlations associated with multiple natural frequencies occurring for the plate both with and without elastic foundations; moreover, for the plate with elastic foundation, the additional cross-correlations due to closeness of (noncoincident) frequencies may become important. The latter case was encountered in the problem of the beam on elastic foundation (section on Crandall's problem), whereas the plate has additional feature of coincident natural frequencies.

chapter 11

Monte Carlo Method

PROBLEM 11.1

Simulate a random variable with Weibull distribution, using the inverse function method.

SOLUTION 11.1

The random variable with Weibull distribution has a density

$$f_Y(y) = \begin{cases} 0, & x < 0 \\ \alpha\beta y^{\beta-1} e^{-\alpha y^{\beta}}, & x > 0 \end{cases}$$

(see Eq. 4.11). The appropriate distribution function is

$$F_Y(Y) = 1 - \exp(-\alpha y^{\beta})$$

which can be easily checked by differentiation. In order to simulate Y through the inverse function method (section 11.3, pp. 438–439) we have to solve

$$F_Y(Y) = X$$

where X has uniform distribution in $(0, 1)$. Or

$$1 - \exp(-\alpha Y^{\beta}) = X$$

yielding

$$\exp(-\alpha Y^{\beta}) = 1 - X$$

$$-\alpha Y^{\beta} = \ln(1 - X)$$

$$Y = \left[-\frac{1}{\alpha}\ln(1 - X)\right]^{1/\beta}$$

or since $X = 1$ and $1 - X$ have identical distribution,

$$Y = \left(-\frac{1}{\alpha}lnX\right)^{1/\beta}$$

PROBLEM 11.2

Solve Problem 5.13 using the Monte Carlo method, and compare with the exact solution.

SOLUTION 11.2

For the probability density function of $\bar{\xi}$ we have

$$f_{\bar{X}}(\bar{\xi}) = be^{-b\bar{\xi}}$$

or

$$f_{\bar{X}}(\bar{\xi}) = 1 - e^{-b\bar{\xi}} = u$$

where u is a realization (possible value) of uniformly distributed random variable, denoted by U. Then equation

$$F_{\bar{X}}(X) = U$$

yields an equation for X, namely

$$X = F_{X}^{-1}(U) = \frac{1}{b}\ln(1 - U)$$

since however, U and $1 - U$ have identical distribution, we use

$$X = \frac{1}{b}lnU$$

We assume, $b = 1$, so that

$$X = -lnU$$

We will simulate U, using formula

$$Y_{i+1} = \alpha y_i \mathrm{mod}(m)$$

with

$$u_i = y_i/m$$

we take, following Rubinstein (p. 23)

$$a = 2^7 + 1 = 129$$

$$m = 2^{14} = 16,384$$

$$y_0 = 1,984$$

Let us estimate the reliability at the level $\alpha = 0.8$ through only 10 realizations, by hand simulation. We find first values of y_i, then $u_i = y_i/m$; initial imperfections become $\bar{\xi} = -lnu_i$. These values of imperfection are used for calculating the buckling loads via formula, following from Eq. 5.53.

$$\frac{\lambda_s}{\lambda_c} = 1 - 2a\bar{\xi} - \sqrt{(1 - 2a\bar{\xi})^2 - 1}$$

Results are summarized in the following table:

y_i	u_i	$\bar{\xi}_i$	$\lambda_s^{(i)}/\lambda_c$
16,896	0.636	0.452	0.284
1,664	0.063	2.764	0.072
2,176	0.082	2.501	0.084
15,104	0.569	0.564	0.250
9,536	0.359	1.024	0.169
8.384	0.316	1.152	0.155
19,136	0.720	0.329	0.336
25,024	0.942	0.060	0.616
14,336	0.540	0.616	0.237
16,704	0.629	0.464	0.280

Now, at level $\alpha = 0.1$,

$$\tilde{R}(0.1) = \mathrm{Prob}(\Lambda > 0.1)$$

we see that only in two instances the system buckled below load level 0.1, implying that the reliability estimate is $(10 - 2)/10 = 0.8$.

Exact solution is

$$R = 1 - \exp\left[-\frac{(1-\alpha)^2}{4(-a)\alpha}\right] = 0.868$$

Noting that we had only 10 realizations in our paper and pencil Monte Carlo Method, the comparison is remarkable.

At level $\alpha = 0.2$,

$$\tilde{R}(0.2) = \text{Prob}(\Lambda > 0.2) = \frac{6}{10}$$

versus exact solution ($a = -1$):

$$R = 1 - \exp\left[-\frac{(1-0.2)^2}{4(-a) \times 0.2}\right] = 0.551$$

again we observe a good comparison for such a small sample.

PROBLEM 11.3

Consider the random vector $\{Y\} = (Y_1, Y_2)$ with joint probability density $f(y_1, y_2) = 5y_1$, for $0 < y_1 < 1, 0 < y_2 < 1$. Simulate this random vector.

SOLUTION 11.3

Note to lecturer: $f_{Y_1Y_2}(y_1, y_2) = 2y_1$

$$f_{Y_1}(y_1) = \int_0^1 f_{Y_1Y_2}(y_1, y_2)dy_2 = \int_0^1 2y_1 dy_2 = 2y_1$$

$$F_{Y_1}(y_1) = \int_0^{y_1} f_{Y_1}(y_1)dy_1 = \int_0^{y_1} 2y_1 dy_1 = \frac{2}{2}y_1^2 = y_1^2$$

$$f_{Y_2}(y_2|y_1) = \frac{f(y_1, y_2)}{f(y_1)} = \frac{2y_1}{2y_1} = 1$$

and

$$F_{Y_2}(y_2|y_1) = \int_0^{y_2} f(y_2|y_1)dy_2 = y_2$$

Therefore, from

$$x_1 = y_1^2$$

$$x_2 = y_2$$

We get our simulation formulas

$$y_1 = \sqrt{x}_1$$

$$Y_2 = X_2$$

PROBLEM 11.4

Perform the Cholesky decomposition of the variance-covariance matrix $v_{jk} = \sigma^{|j-k|}$, where σ is a positive number. Simulate a normal random vector with zero mathematical expectation and this variance-covariance matrix.

SOLUTION 11.4

According to Equations (11.19), we have

$$C_{j1} = \frac{v_{j1}}{\sqrt{v_{11}}} = \frac{\sigma^{|j-1|}}{\sqrt{\sigma}^{|1-1|}} = \sigma^{|j-1|}$$

$$C_{jj} = \left(v_{jj} - \sum_{i=1}^{j-1} c_{ji}^2 \right)^{1/2}, \quad 1 \leq j \leq N$$

Now

$$v_{jj} = 1$$

$$\sum_{i=1}^{j-1} c_{ji}^2 = \sum_{i=1}^{j-1} \sigma^{2|j-i|}$$

But since $i < j$, we have

$$\sum_{i=1}^{j-1} \sigma^2 |j-i| = \sum_{i-1}^{j-1} \sigma^{2(j-i)} = \sigma^{2j} \sum_{i=1}^{j-1} (\sigma^{-2})^i$$

$$= \sigma^{2j} \frac{(\sigma^{-2})^j - \sigma^{-2}}{\sigma^{-2} - 1} = \frac{1 - \sigma^{2(j-1)}}{\sigma^{-2} - 1}$$

and

$$C_{jj} = \left(1 - \frac{1 - \sigma^{2(j-1)}}{\sigma^{-2} - 1} \right)^{1/2}$$

Now

$$C_{jk} = \frac{v_{jk} - \sum_{i=1}^{k-1} C_{ji} C_{ki}}{C_{kk}}, \quad 1 < k < j \leq N$$

Now

$$\sum_{i=1}^{k-1} C_{ji}C_{ki} = \sum_{i=1}^{k-1} \sigma^{|j-i|}\sigma^{|k-i|}$$

but since $i < k,\ k < j$, hence $i < j$ and

$$\sum_{i=1}^{k-1} \sigma^{|j-i|}\sigma^{|k-i|} = \sum_{i=1}^{k-1} \sigma^{j-i-(k-i)} = \sum_{i=1}^{k-i} \sigma^{j-k-2i}$$

$$= \sigma^{j-k} \sum_{i=1}^{k-1} (\sigma^{-2})^i = \sigma^{j-k} \frac{(\sigma^{-2})^k - \sigma^{-2}}{\sigma^{-2} - 1}$$

$$= \frac{\sigma^{j-3k} - \sigma^{j-k-2}}{\sigma^{-2} - 1}$$

and

$$C_{jk} = \frac{\sigma^{|j-k|} - (\sigma^{j-3k} - \sigma^{j-k-2})(\sigma^{-2} - 1)^{-1}}{\{1 - [1 - \sigma^{2(j-1)}][\sigma^{-2} - 1]^{-1}\}^{1/2}}, \quad 1 < k < j \le N.$$

i	Z_1^*	Z_2^{**}	P_1	P_2	$\lvert M_1\rvert \cdot 10^{-3}$	$\lvert M_2\rvert \cdot 10^{-3}$	Z_3^{***}	σ_Y	$[M_y = \sigma_y \cdot S] \times 10^{-3}$	Conclusion
1	0.284	−1.016	56.8	−295.2	181.6	647.2	0.94557	294.6	530.3	fails
2	0.458	0.360	91.7	216.4	399.8	524.5	0.28573	228.6	411.5	fails
3	1.307	−0.119	261.4	220.2	743	701.8	0.67897	267.9	482.3	fails
4	−1.625	2.331	−325	482.5	167.5	817.5	0.54387	254.4	457.9	fails
5	−0.629	1.672	−125.8	453.4	201.8	781.0	0.54622	254.6	458.3	fails
6	−0.504	1.053	−100.8	−465.6	667.2	1032.0	0.64431	244.4	439.9	fails
7	−0.056	0.840	−11.2	279.8	257.4	548.4	0.91190	291.2	524.6	fails
8	−0.131	−0.246	−26.2	−111.4	163.8	249.0	0.42592	242.6	242.6	does not fail
9	0.048	0.237	9.6	91.7	110.9	193.0	0.92927	292.6	527.2	does not fail
10	1.879	−1.312	375.8	−78.3	672.9	219.2	0.45973	246.0	442.8	fails

*row no. 0025 (see page 317 of the Manual)
**row no. 0026 (see page 317 of the Manual)
***row no. 00027 (see page 316 of the Manual)

$Z_3 = 0 \cdot E_1E_2E_3E_4E_5$ where E_j are independent random digits (see Eq. 11.4 of the text).

The Monte Carlo estimate of the reliability turns out to be

$$\tilde{R} = \frac{2}{10} = 0.2.$$

PROBLEM 11.5

A beam simply supported at its ends is subjected at sections $x = l/3$ *and* $x = 2l/3$ (where l is the span of the beam) to two concentrated loads with identically vanishing mean values and the variance-covariance matrix.

$$[V] = \sigma^2 \begin{bmatrix} 1 & 1 \\ 1 & 4 \end{bmatrix}$$

Show that the matrix $[C]$ is

$$[C] = \sigma \begin{bmatrix} 1 & 0 \\ 1 & \sqrt{3} \end{bmatrix}$$

Determine the reliability of the beam by the Monte Carlo method given that the allowable stress is also a random, variable uniformly distributed in the interval (assuming actual values to l, σ, c and the section modulus S of the beams).

SOLUTION 11.5

It is straightforward to show that

$$[C][C]^T = [V]$$

This can be checked by direct multiplication of matrices $[C]$ and $[C]^{\mathrm{T}}$.

Realization of the concentrated loads are given by equation 11.24

$$Y = [C]\{Z\} + \{m\}$$

or

$$Y_1 = \sigma Z_1 + m_1$$

$$Y_2 = \sigma(Z_1 + \sqrt{3}Z_2) + m_2$$

where m_1 and m_2 are mathematical expectations of loads Y_1 and Y_2 respectively, Z_1 and Z_2 are independent standard normal variables.

Once the values of length ℓ, σ, m_1 and m_2 are chosen, the maximum stresses should be calculated as they occur in cross sections where the loads Y_1 and Y_2 are applied. The stresses should be compared with allowable stress, stemming from simulation of uniformly distributed random variable in the interval $[\sigma_{Y,L}, \sigma_{Y,U}]$ where $\sigma_{Y,L}$ is the lower bound of the yield stress and $\sigma_{Y,U}$ is the upper bound of it. If the maximum stress in either cross section exceeds $\sigma_{Y,L}$ the beam has failed. We have to count the number of failed beams from the fixed total of beams N. It is recommended to fix N at 10^6. For the selected values of constants, please calculate how many beams will fail. What is the estimated reliability?

PROBLEM 11.6

What is the reliability in Problem 11.5, if the variance-covariance matrix is

$$[V] = \sigma^2 \begin{bmatrix} 1 & 2-\varepsilon \\ 2-\varepsilon & 4 \end{bmatrix}$$

Investigate the variation of the reliability in the interval $0 < \varepsilon \leq 1$, checking the case $\varepsilon = 0$ separately.

SOLUTION 11.6

For $\varepsilon = 0$, the matrix $[V]$ becomes

$$[V] = \sigma^2 \begin{bmatrix} 1 & 2 \\ 2 & 4 \end{bmatrix}$$

and the lower-triangular matrix $[C]$ is

$$[C] = \sigma \begin{bmatrix} 1 & 0 \\ 2 & 0 \end{bmatrix}$$

Therefore, loads P_1 and P_2 are simulated as follows

$$\begin{Bmatrix} P_1 \\ P_2 \end{Bmatrix} = 200 \begin{bmatrix} 1 & 0 \\ 2 & 0 \end{bmatrix} \begin{Bmatrix} Z_1 \\ Z_2 \end{Bmatrix}$$

Instead of the table in Problem 11.5 we get

i	Z_1^*	P_1	$P_2 = 2P_1$	$\lvert M_1 \rvert \cdot 10^{-3} = 4P_1 \times 10^{-3}$	$\lvert M_2 \rvert \cdot 10^{-3}$	Z_3^{**}	σ_y	$[M_y = \sigma_y S] \times 10^{-3}$	Conclusion
1	0.284	56.8	113.6	227.2	284.0	0.94557	294.6	530.3	does not fail
2	0.458	91.7	183.4	366.8	458.5	0.28573	228.6	411.5	fails
3	1.307	261.4	522.8	1045.6	1307.0	0.67897	267.9	482.3	fails
4	−1.625	-325	650	1300	1625	0.54387	254.4	457.9	fails
5	−0.639	−125.8	−251.6	503.2	629.0	0.54622	254.6	458.3	fails
6	−0.504	100.8	−201.6	403.2	504.0	0.64431	244.4	439.9	Fails
7	−0.056	−11.2	−22.4	44.8	56	0.91190	291.2	524.2	does not fail
8	−0.131	−26.2	−52.4	104.8	131	0.42592	242.6	436.7	does not fail
9	0.048	9.6	19.2	34.4	48.0	0.92927	292.6	527.2	does not fail
10	1.879	375.8	751.6	1503.2	1879.0	0.45973	246.0	442.8	fails

*For realizations of Z_1 see page 317 of the Manual (row 0025).

**$Z_3 = 0.E_1E_2E_3E_4E_5$, where E_j are independent random digits (see Eq. 11.4 of the text). For the realizations of E_j see page 316 of the Manual.

Z_1^*	Z_2^*	Y_1	Y_2	$\|M_1\|$	$\|M_2\|$	Conclusion
0.284	−1.016	1,585.2	1,454.6	388,137.5	371,812.5	does not fail
0.4586	0.360	1,637.6	1,600.0	407,050.0	402,350.0	does not fail
1.307	−0.199	1,892.1	1,685.8	460,131.2	434,343.8	Fails
−1.625	2.331	1,012.5	1,458.1	280,975.0	336,675.0	does not fail
−0.629	1.672	1,311.3	1,550.4	342,775.0	372,581.2	does not fail
−0.504	1.053	1,348.8	1,515.6	347,625.0	368,475.0	does not fail
−0.056	0.840	1,483.2	1,564.2	375,862.5	385,987.5	does not fail
−0.131	−0.246	1,460.7	1,459.1	365,075.0	364,875.0	does not fail
0.048	0.237	1,514.4	1,527.7	379,431.8	381,093.8	does not fail
1.879	−1.312	2,063.7	1,331.8	470,181.2	378,693.8	fails

Now, yielding moment is

$$M_Y = \sigma_Y \cdot S = 250 \times 1800 = 450,000\, Nmm$$

This means (see the table) that eight beams out of ten are not failing, and therefore the estimate of the reliability is 0.8.

Exact solution yields 0.806.

PROBLEM 11.7

A beam simply supported at its ends is subjected to two concentrated loads Y_1 and Y_2 acting at sections $x = l/4$ and $x = 3l/4$, respectively, $l = 1000$ mm, $S = 1800\,\text{mm}^3$, $E(Y_1) = 1500$ N, yield sstress $= 250\,\text{N/mm}^2$. The variance-covariance matrix is (loads are normally distributed).

$$[V] = 300^2 \begin{bmatrix} 1 & \frac{1}{2} \\ \frac{1}{2} & \frac{1}{3} \end{bmatrix} N^2$$

Find the reliability of the beam both exactly and by the Monte Carlo method.

SOLUTION 11.7

This evaluation of the matrices $[C]$ can be done via Cholesky decomposition formula (11.19) on page 444 of the textbook. However this two-dimensional problem can be solved directly by assuming the form of matrix $[C]$ as follows

$$[C] = 300 \begin{bmatrix} 1 & 0 \\ \alpha & \beta \end{bmatrix}$$

The value 300 is obtained since the transposed matrix $[C]^{\mathrm{T}}$ contains 300 as a factor which will lead to the factor 300^2 for variance-covariance matrix $[V]$ as given in the problem. Then we get

$$[C][C]^T = 300\begin{bmatrix} 1 & 0 \\ \alpha & \beta \end{bmatrix} 300\begin{bmatrix} 1 & \alpha \\ 0 & \beta \end{bmatrix}$$

$$= 300^2\begin{bmatrix} 1 & \alpha \\ \alpha & \alpha^2+\beta^2 \end{bmatrix}$$

Comparing this result with matrix $[V]$

$$[V] = 300^2\begin{bmatrix} 1 & \frac{1}{2} \\ \frac{1}{2} & \frac{1}{3} \end{bmatrix}$$

we obtain

$$\alpha = \frac{1}{2}$$

Thus

$$\alpha^2 + \beta^2 = \frac{1}{3}$$

or

$$\frac{1}{4} + \beta^2 = \frac{1}{3}$$

we find value of β as

$$\beta = \sqrt{\frac{1}{3} - \frac{1}{4}} = \frac{1}{\sqrt{12}} = \frac{1}{2\sqrt{3}}$$

as a result matrix $[C]$ gets the following form

$$[C] = 300\begin{bmatrix} 1 & 0 \\ \frac{1}{2} & \frac{1}{2\sqrt{3}} \end{bmatrix}$$

Therefore, the realizations Y_1 and Y_2 of concentrated loads are obtained as follows

$$Y_1 = 300Z_1$$

$$Y_2 = 150Z_1 + \frac{150}{\sqrt{3}}Z_2$$

Note that this result coincides with the $[C]$ matrix obtainable from equations between (11.24) and (11.25) on page 446 of the textbook, if we limit ourselves only with elements C_{11}, C_{12}, C_{21} and C_{22}. We get C_{11} =1, C_{12} =0, $C_{21} = \frac{1}{2}$ and $C_{22} = \frac{\sqrt{3}}{6} = \frac{1}{2\sqrt{3}}$. Further steps of the solution are straightforward.

Note that the Monte Carlo Method involves simulating thousands of beams and counting the number of beams that do not fail, forv estimating the beam's reliability. The somewhat nontrivial part is randomly generating said simulations. For this analysis, MATLAB's randn () function was used to create randomly generated numbers in a normal distribution. The *rand ()* function is a pseudorandom number generator that uses Mersenne Twister that has a period of $2^{19937} - 1$, which is sufficiently long enough for this case.

Consider a slightly more complicated case than given by the problem 11.7. The randn() function creates a normal distribution around the mean zero and with a variance of unity. The method of linear transformation is then used along with the provided mean vector and variance-covariance matrix to generate a number of loading simulations. The results of the simulations are used to tally non-failures to estimate reliability. MATLAB was used for all calculations for the analysis that will follow. The reliability estimation was conducted by M. Kothawala.

CASE 1: SIMPLY SUPPORTED BEAM

A graphical representation of the problem is shown below:

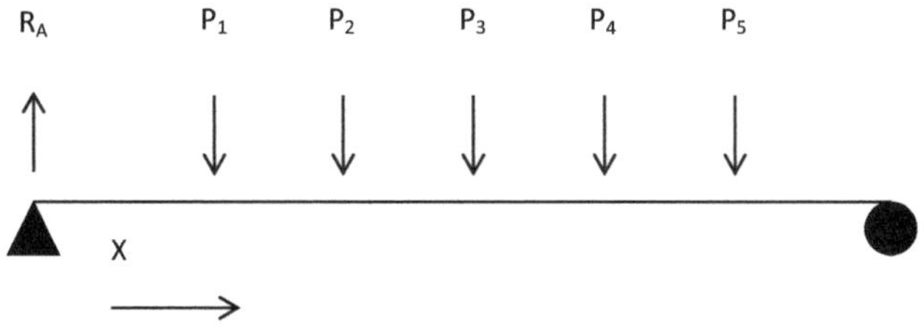

Beam parameters are listed below:

Table I: Beam parameters

Length	L	1000 mm
Allowable Stress	σ_y	250 N/mm^2
Section Modulus	S	1800 mm^3

Load locations, the mean vector of forces, and the variance covariance matrix are listed below:

Table II: Load positions on beam (from left end of beam)

P1	300 mm
P2	400 mm
P3	500 mm
P4	600 mm
P5	700 mm

$$M^T = [800\,400\,400\,800\,400]$$

$$[V] = 100^2 \begin{bmatrix} 1 & 1/2 & 1/3 & 1/4 & 1/5 \\ 1/2 & 1/3 & 1/4 & 1/5 & 1/6 \\ 1/3 & 1/4 & 1/5 & 1/6 & 1/7 \\ 1/4 & 1/5 & 1/6 & 1/7 & 1/8 \\ 1/5 & 1/6 & 1/7 & 1/8 & 1/9 \end{bmatrix}$$

The mean vector and covariance-variance matrix are modeled after Example 11.3 mentioned above. The reaction at the left hand support is

$$R_A = 0.1(7P_1 + 6P_2 + 5P_3 + 4P_4 + 3P_5)$$

The maximum stress is where the maximum bending moment is located. The expression for the bending moment reads

$$M(x) = -0.1(7P_1 + 6P_2 + 5P_3 + 4P_4 + 3P_5)x + P_1\langle x - 0.3L\rangle$$
$$+P_2\langle x - 0.4L\rangle + P_3\langle x - 0.5L\rangle + P_4\langle x - 0.6L\rangle + P_5\langle x - 0.7L\rangle$$

The location of maximum bending moment is at the location of the largest load. Since the loads are random variables and the largest load is unknown, the moment at each load location needs to be calculated. The bending moments in the cross sections with the loads are

$$M_1 = M(.3L) = -0.01L(21P_1 + 18P_2 + 15P_3 + 12P_4 + 9P_5)$$

$$M_2 = M(.4L) = -0.01L(18P_1 + 24P_2 + 20P_3 + 16P_4 + 12P_5)$$

$$M_3 = M(.5L) = -0.01L(15P_1 + 20P_2 + 25P_3 + 20P_4 + 15P_5)$$

$$M_4 = M(.6L) = -0.01L(12P_1 + 16P_2 + 20P_3 + 24P_4 + 18P_5)$$

$$M_5 = M(.7L) = -0.01L(9P_1 + 12P_2 + 15P_3 + 18P_4 + 21P_5)$$

Stress is then calculated per the following equation

$$\sigma_i = M_i/S$$

where i denotes the location of the stress. The reliability is then the probability that the absolute value of the normal stress does not exceed the yield stress in any of the five locations.

The Cholesky decomposition yields the lower triangular matrix $[C]$:

$$[C] = 100\begin{bmatrix} 1 & 0 & 0 & 0 & 0 \\ .5 & 28.8675 & 0 & 0 & 0 \\ 33.33 & 28.8675 & 7.4536 & 0 & 0 \\ .25 & 25.9808 & 11.1803 & 1.8898 & 0 \\ 20 & 23.0940 & 12.7775 & 3.7796 & 0.4762 \end{bmatrix}$$

The vector $\{Y\}$ is then simulated using the following equation:

$$\{Y\} = [C]\{V\} + [m]$$

The elements of vector Z were calculated using MATLAB's *randn()* function. In ten series of 100,000, nonfailures totaled 6913, 6700, 6832, 6930, 6881, 6836, 6898, 6871, 6821, 6923, so that the reliability estimate of 1,00,000 beams is $R* = 0.0686$.

CASE 2: CANTILEVER BEAM

A graphical representation of the problem is shown below:

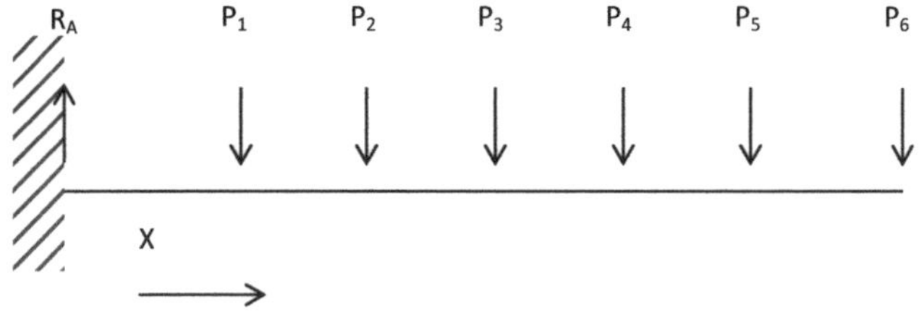

Beam parameters are listed below:

Table III: Beam parameters

Length	L	1000 mm
Yield Stress	σ_y	250 N/mm^2
Section Modulus	S	1800 mm^3

Load locations, the mean vector of forces, and the variance covariance matrix are listed below:

Table IV: Load positions on beam (from left end of beam)

P1	300 mm
P2	400 mm
P3	500 mm
P4	600 mm
P5	700 mm
P6	1000 mm

$$M^T = [200\,100\,100\,200\,100\,200]$$

$$[V] = 100^2 \begin{bmatrix} 1 & 1/2 & 1/3 & 1/4 & 1/5 & 1/6 \\ 1/2 & 1/3 & 1/4 & 1/5 & 1/6 & 1/7 \\ 1/3 & 1/4 & 1/5 & 1/6 & 1/7 & 1/8 \\ 1/4 & 1/5 & 1/6 & 1/7 & 1/8 & 1/9 \\ 1/5 & 1/6 & 1/7 & 1/8 & 1/9 & 1/10 \\ 1/6 & 1/7 & 1/8 & 1/9 & 1/10 & 1/11 \end{bmatrix}$$

The covariance-variance matrix is modeled after Example 11.3 in the textbook. The mean vector was also modeled after Example 11.3 at first but results led to zero reliability. The values of the elements of the mean vector were then quartered to produce a more interesting result.

The maximum stress is located at the location of the maximum moment, which is at root (left hand edge in this case) of a cantilever beam. The equation for the maximum moment is

$$M_{max} = 300P_1 + 400P_2 + 500P_3 + 600P_4 + 700P_5 + 1000P_6$$

Stress is then calculated per the following equation

$$\sigma = M/S$$

The reliability is then the probability that the absolute value of the normal stress does not exceed the allowable stress.

The method of linear transformation was used to generate the 100,000 normally distributed values for each force. The Cholesky decomposition yields the lower triangular matrix [C]:

$$[C] = 100 \begin{bmatrix} 1 & 0 & 0 & 0 & 0 & 0 \\ .5 & 28.8675 & 0 & 0 & 0 & 0 \\ 33.3333 & 28.8675 & 7.4536 & 0 & 0 & 0 \\ .25 & 25.9808 & 11.1803 & 1.8898 & 0 & 0 \\ 20 & 23.0940 & 12.7775 & 3.7796 & 0.4762 & 0 \\ 16.6667 & 20.6197 & 13.3099 & 5.2495 & 1.1905 & 0.1196 \end{bmatrix}$$

The vector $\{Y\}$ is then simulated using the following equation:

$$\{Y\} = [C]\{V\} + [m]$$

The elements of vector Z were calculated using MATLAB's *randn()* function. In ten series of 100,000, non-failures totaled 26278, 26142, 26040, 26178, 26345, 26073, 26156, 26282, 26221, 26201, so that the reliability estimate R^* for 1,000,000 beams turns out to be $R* = 0.2419$.

PROBLEM 11.8

Consider a bar on a mixed quadratic-cubic elastic foundation, with its displacement described by the differential equation

$$EI\frac{d^4w}{dx^4} + P\frac{d^2w}{dx^2} + k_1w - k_2w^2 - k_3w^3 = -P\frac{d^2\bar{w}}{dx^2}$$

where the constants k_1, k_2, and k_3 are the foundation parameters. Using the Galerkin method in a single-term approximation, derive an equation analogous to Eq. (11.69) and draw a parallel to the model structure considered in Sec. 5.5. Show also that the relation between the nondimensional buckling load α^* and the initial imperfection $\bar{\xi}_m$ is in full agreement with Eq. (5.50). Finally, derive the reliability function and the allowable nondimensional load. (Multiterm analysis of this problem is given by Elishakoff, 1981.)

SOLUTION 11.8

Consult with the first part of the paper entitled "Reliability Approach to the Random Imperfection Sensitivity of Columns", by I. Elishakoff, Acta Mechanica, Vol. 58, pp. 151–170, 1985.

Table of 2500 Realizations of a Random Digit [taken from "A Million Random Digits with 100,000 Normal Deviates", by Rand Corporation, The Free Press]

Row number	Realizations (group)									
00025	61196	90446	26457	47774	51924	33729	65394	59593	42582	60527
00026	15474	45266	95270	79953	59367	83848	82396	10118	33211	59466
00027	94557	28573	67897	54387	54622	44431	91190	42592	92927	45973
00028	42481	16213	97344	08721	16868	48767	03071	12059	25701	46670
00029	23523	78317	73208	89837	68935	91416	26252	29663	05522	82562

Table of Fifty Realizations of a Normal Random Variable $N(0, 1)$ [taken from "A Million Random Digits with 100,000 Normal Deviates", by Rand Corporation, The Free Press]

Row number	Realizations									
0025	0.284	0.458	1.307	−1.625	−0.629	−0.504	−0.056	−0.131	0.048	1.879
0026	−1.016	0.360	−0.119	2.331	1.672	−1.053	0.840	−0.246	0.237	−1.312
0027	1.603	−0.952	−0.566	1.600	0.465	1.951	0.110	0.251	0.116	−0.957
0028	−0.190	1.479	−0.986	1.249	1.934	0.070	−1.358	−1.246	−0.959	−1.297
0029	−0.722	0.925	0.783	−0.402	0.619	1.826	1.272	−0.945	0.494	0.050

chapter 12

Additional Example Problems and Comments

In this chapter, we provide additional example problems and comments on examples in the textbook. These examples are designed so as to elucidate some subtle points, and draw attention to unanticipated twist in the obtained results.

Example 2.8

Comment on Example 2.8 (p 35)

At the end of the problem, the following question can be put to the students:

Could you anticipate the result that the probability of a white ball being removed from the second box is unaffected by adding the ball from the first box?

Some students come out with the following idea: Let us visualize a string with a mass distribution such that the entire mass equals unity and with length such that the mass density per unit length is given by the probability of drawing a white ball, $a/(a+b)$. Now another string is taken with the same mass density and length (analogous to another box with "a" white balls and "b" black balls). Now if we cut off a piece from the string and add it to the second, the mass density of the resulting enlarged string will remain the same.

Additional Example

In connection to Example 2.8 consider the following problem.

A box contains one call (call it ball No. 1) which is equally likely to be white or black. A white ball (call it ball No. 2) is placed in the box and a ball is removed at random from the box. Find the probability of

"the remaining ball is white"

SOLUTION

One way to solve this problem is to formulate it as a problem of drawing successively two objects from a box. The objects to be drawn from the box are:

$W_1 \to$ white ball (No. 1)
$B_1 \to$ black ball (No. 1)
$W_2 \to$ white ball (No. 2) (or simply ball No. 2)

Note: there is no B_2.

Now the sample space consists of four elementary events (sample points) and is shown in the figure below.

$A_1 = (W_1, W_2)$	$A_2 = (W_2, W_1)$
$A_3 = (W_2, B_1)$	$A_4 = (B_1, W_2)$

where for example the pair (W_1, W_2) means the first draw is white ball (No. 1) and the second draw is white ball (No. 2); i.e. ball No. 2. The other pairs are defined alike. It is clear that $P(A_1) = P(A_2) = P(A_3) = P(A_4) = \frac{1}{4}$

In fact, by the total probability theorem

$$P(A_1) = P(A_1|\{1\})P\{1\}) + P(A_1|\{2\})P(\{2\})$$

where $\{j\}$ means jth ball drawn. But, since $A_1 \subset \{1\}$, $P(A_1|\{2\})$ vanishes and $P(A_1) = P(A_1|\{1\})P(\{1\}) = \frac{1}{4}$ *QED*. Let A and B be the following events

A – the first draw is white
B – the second draw is white. Then,
$A = A_1 \cup A_2 \cup A_3, \quad B = A_1 \cup A_2 \cup A_4$

Now, in view of the above formulation the required probability is the conditional probability given by:

$$P\{\text{white ball is left in the box}\} = P(B|A) = \frac{P(B\cap A)}{P(A)}$$

$$= \frac{P(A_1 \cup A_2)}{P(A_1 \cup A_2 \cup A_3)}$$

$$= \frac{P(A_1) + P(A_2)}{P(A_1) + P(A_2) + P(A_3)} = \frac{\frac{1}{4}}{\frac{3}{4}} = \frac{2}{3}$$

Additional remarks, Prob. 2.6

[reproduced Form Emile Borel, Probabilité et Certitude, Presses Universitaires de France, Paris, 1956].

All civilized countries have for more than a century now, have been practicing registration of birth with their sex duly noted, errors being extremely rare. By this means, statistics for a million birth can be compiled over a period ranging from several months to several years, depending on the size and fertility of the country's population. It turns out regularly that the number of boys exceeds 510,000 and that of girls is less than 490,000. It can thus be concluded that these countries, for more than a century, the probability of boys being born is slightly in excess of 1/2, close to 0.51. This is a very interesting result from the biological point of view. Indeed, modern theories of heredity conclude that the sex of the child is determined at the moment of conception. The first question is then, whether the probabilities for the two sexes are equal from this moment, or alternatively whether the observed difference is due to a higher mortality rate of female fetuses in utero or to the proportion of male fetuses being higher already from the moment of conception.

Note that the figure 0.51 means 1040 boys versus 1000 girls. In fact, the number of male births in France versus 1000 females births (only live births taken into account) is illustrated by the following statistics:

1924/25	1,050
1935/39	1,038
1940	1,040
1941	1,038
1942	1,055
1943	1,061
1944	1,059
1945	1,056
1946	1,056
1947	1,057

Interestingly, the proportion of male births is seen to increase slightly following major wars.

It is also remarkable, that according to recent research by an Australian genetist, the ratio of boys to girls among the children of sailors is 2:1, while among those of pilots girls predominate. This latter phenomenon was observed also among pilots of the Israeli Airlines, although the volume of data and observation span might be insufficient for reliable conclusions as to any influence of the father's profession on the child's sex.

We assume in problem 2.6 that the probabilities of male and female births are equal.

Additional Example, Section 4.4: Poisson Distribution

Optimization Example

So long as it performs successfully, a mechanical device yields a profit of $\alpha > 2$ dollars per week to a company. It has, however, occasional failures, whose number over a period of t weeks, is a Poisson-distributed random variable with parameter t, i.e.

$$P(X = m) = \frac{t^m}{m!}e^{-t}, \quad m = 1, 2, \ldots$$

Choose t so that the mean profit be maximum, given that the loss incurred through the failures is $(X^3 + X)$ dollars.

SOLUTION

The total profit over any period of t weeks is

$$P = \alpha t - (X^3 + X)$$

The mean profit is

$$E(P) = \alpha t - E(X^3 + X)$$

Let us calculate $E(X^3)$. According to the next-to last equation on p. 60, we have

$$\left[\frac{d^3\psi_X(\theta)}{d\theta^3}\right]_{\theta=0} = -\frac{m_3 - 3m_1m_2 + 2m_1^3}{i^3}$$

Now, as given on p. 75, for a Poisson distribution,

$$\psi_X(\theta) = te^{i\theta} - t$$

With the parameter "a" being replaced here by t, in accordance with the conditions of the problem:

$$\left[\frac{d^3\psi_X(\theta)}{d\theta^3}\right]_{\theta=0} = -it$$

and

$$m_3 = E\left(X^3\right) = 3m_1m_2 - 2m_1^3 - i^3\left[\frac{d^3\psi_X(\theta)}{d\theta^3}\right]_{\theta=0}$$

$$= 3t\left(t+t^2\right) - 2t^3 + t$$

since

$$m_2 = E(X^2) = \text{Var}(X) + [E(X)]^2 = t + t^2$$

Finally

$$E(X^3) = t^3 + 3t^2 + t$$

Hence, the mean profit is

$$E(P) = \alpha t - (t^3 + 3t^2 + t + t)$$

To find the maximum, we equate the first derivative of $E(P)$ to zero,

$$\alpha - 3t^2 - 6t - 2 = 0$$

or

$$3t^2 + 6t + (2 - \alpha) = 0$$

Whence

$$t_* = \frac{-6 \pm \sqrt{36 - 12(2-\alpha)}}{6} = \frac{-6 \pm \sqrt{12 + 12\alpha}}{6} = \frac{-3 \pm \sqrt{3(1+\alpha)}}{3}$$

For example for $\alpha = 1,199$, we have

$$t_* = \frac{-3 + \sqrt{3 \cdot 1200}}{3} = 19 \text{ weeks.}$$

Now, the maximum mean profit is

$E(P) = 1199 \cdot 19 - (19^3 + 3 \cdot 19^2 + 2 \cdot 19) = \$14,801$ in 19 weeks. [Note that α should be larger than 2. Indeed for $\alpha = 2$, $t_* = 0$, and performance time is zero. No profit can be made however without work: The "maximum" profit $E(P)$ corresponding to $t_* = 0$ is zero too!]

Additional Examples, Section 4.13:

Moments of a Function of a Random Variable (pg. 85)

Example 1.

The radius R of a circle is a random variable with given probability density $f_R(r)$. Find the average area, $E(A)$

In accordance with Eq. (4.26), for $k = 1$

$$m_1(Y) = E(Y) = \int_{-\infty}^{\infty} \phi(x) f_X(x) dr$$

In our case, $Y \equiv A,\ X \equiv R,\ \phi(x) = \pi r^2$, and

$$E(A) = \int_{-\infty}^{\infty} \pi \zeta^2 f_R(r) dr = \int_{0}^{\infty} \pi r^2 f_R(r) dr$$

since r can take on only positive values. For R, uniformly distributed between $(0, c)$:

$$E(A) = \int_0^c \pi r^2 \frac{1}{c} dr = \frac{\pi}{c}\left[\frac{r^3}{3}\right]_0^c = \frac{\pi c^2}{3}$$

Note that the area of a circle with mean radius $c/2$ will be

$$\pi [E(R)]^2 = \pi \left(\frac{c}{2}\right)^2 = \frac{\pi c^2}{4}$$

So that replacement of the mean area $\pi c^2/3$ by the latter value implies an error of

$$n = \frac{\frac{\pi c^2}{3} - \frac{\pi c^2}{4}}{\frac{\pi c^2}{3}} 100\% = \frac{1/12}{1/3} 100\% = 25\%$$

Example 2.

Care should be taken not to confuse

$$E(1/X) \text{ with } 1/E(X).$$

Indeed, consider a bar under a tensile deterministic load, $= 1\,kN$; the area X is a random variable. We are interested in the average stress:

$$E\left(\sum\right) = E\left(\frac{1}{X}\right) = \int_{-\infty}^{\infty} \frac{1}{x} f_X(x) dx$$

For X uniformally distributed over $[\alpha, \alpha+\beta,]$, we have

$$f_X(x) = \begin{cases} \frac{1}{\beta}, & \alpha < x < \alpha+\beta \\ 0, & \text{otherwise} \end{cases}$$

and

$$E\left(\sum\right) = \int_{\alpha}^{\alpha+\beta} \frac{1}{x} \cdot \frac{1}{\beta} dx = \frac{1}{\beta} \, lnx \Big]_{\alpha}^{\alpha+\beta} = \frac{1}{\beta} ln \frac{\alpha+\beta}{\alpha}$$

Now

$$\frac{1}{E(X)} = \frac{1}{\alpha+\beta/2}$$

Note that for small β, i.e. for $\beta \gg \alpha$, we have

$$E\left(\sum\right) = \frac{1}{\beta} \ln\left(1+\frac{\beta}{\alpha}\right) \approx \frac{1}{\beta} \cdot \frac{\beta}{\alpha} = \frac{1}{\alpha}$$

whereas also

$$\frac{1}{E(X)} = \frac{1}{\alpha+\frac{\beta}{2}} \approx \frac{2}{\beta}$$

that is, the first tends to infinity, whereas the second remains finite!

Section 5.1

Additional Material to Section 5.1

As is known, the deterministic strength requirement for a bar under a tension

$$\sigma \le \sigma_{allow} \tag{1}$$

gives rise to three basic problems of deterministic strength of materials (DMS).

The first basic problem of DSM is checking the strength requirement (1). For a given bar under tensile forces the axial force $n(x)$ is calculated for any cross section x; the stresses are then determined

$$\sigma(x) = \frac{n(x)}{a(x)}$$

and their value compared with σ_{allow}. If $\sigma(x) \le \sigma_{allow}$, we state that the strength requirement is met, otherwise we state that this requirement is violated.

Example 1.

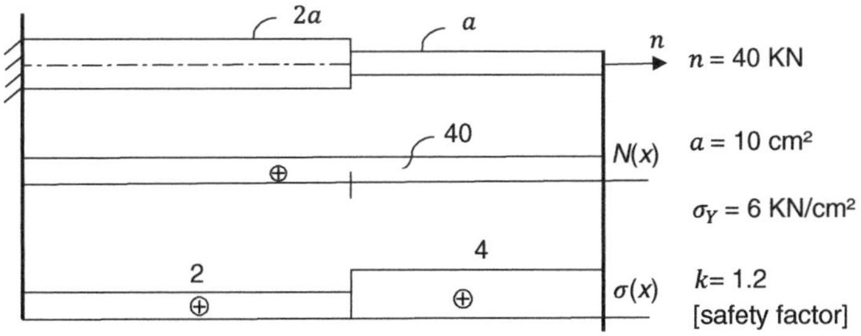

$$n = 40\ KN\ a = 10cm^2\ \sigma_Y = 6KN/cm^2\ k = 1.2\ [safety factor]$$

In this case

$$\sigma\ < \sigma_{allow} = \frac{6}{1.2} = 5\ KN/cm^2$$

and the strength requirement is met.

Let N be proportional to $\sigma_Y a$, such that

$$N = \alpha\sigma_Y a$$

then

$$\sigma_{max} = \frac{\alpha\sigma_Y a}{a} = \alpha\sigma_Y \leq \sigma_{allow} = \frac{\sigma_Y}{k}$$

and the strength requirement is met if

$$\alpha \leq \frac{1}{k}$$

for specified $k = 1.2$, $\alpha \leq 0.8333$, and if $\alpha \leq 8.333$ the strength requirement is met.

The second basic problem of DSM is that of determination of the maximal allowable force parameter. Axial forces are given as proportional to a parameter n. The stress is the cross-section x is determined as $n(x) = nf(x)$ where $f(x)$ is found by the method of cutting of cross-sections, and depends on the position and magnitude of the axial forces (distributed or concentrated). Using the strength requirement inequality (1), we are looking for n_{allow} which would satisfy it. The maximum value $n_{max,allow}$ is obtained when Eq. (1) becomes an equality:

$$n_{max,allow} = \frac{maxf(x)}{A(x)} = \sigma_{allow}$$

and

$$n_{max,allow} = \frac{\sigma_{allow}}{\max[f(x)/A(x)]}$$

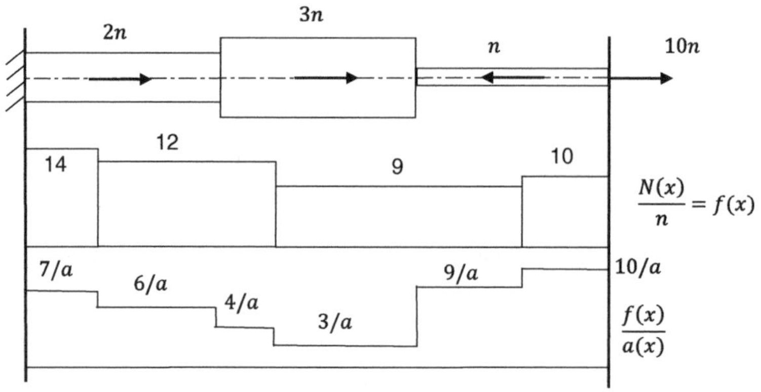

Example 2.

$$max\frac{f(x)}{a(x)} = \frac{10}{a}; \quad n_{max,allow} = \frac{\sigma_{allow}}{10/a} = \frac{a\sigma_{allow}}{10}$$

Now let

$$\sigma_Y = 6KN/cm^2 \; k = 1.2 \,[\text{required factor of safety}] a = 20\,cm^2$$

then

$$\sigma_{allow} = \frac{\sigma_Y}{k} = \frac{6}{1.2} = 5KN/cm^2$$

$$n_{max,allow} = \frac{a \cdot \sigma_{allow}}{10} = \frac{20 \cdot 5}{10} = 10\,KN$$

The third basic problem of DSM is that of design of the bar. The cross-sectioned areas of the bar segments are proportioned to some parameter "a", so that $a(x) = a\phi(x)$. The stress $\sigma(x)$ is again calculated as

$$\sigma(x) = \frac{n(x)}{a(x)} = \frac{n(x)}{a\phi(x)},$$

Where $n(x)$ are determined via the general external forces. Again, the strength requirement is

$$\frac{n(x)}{a\phi(x)} \leq \sigma_{allow}$$

The value "a" which this inequality becomes an equality is the minimum required cross-section area

$$\frac{1}{a_{min,req}} max \frac{n(x)}{\phi(x)} = \sigma_{allow}$$

$$a_{min,req} = \frac{1}{\sigma_{allow}} max \frac{n(x)}{\phi(x)}$$

Example 3.

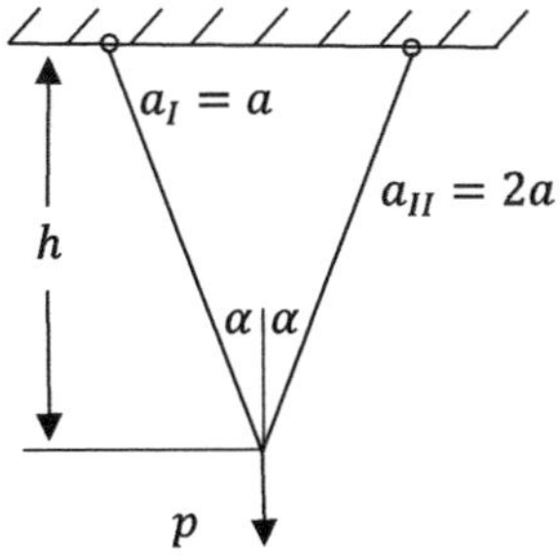

$$p = 120\,KN\;\alpha = 45°\;h = 2m\;\frac{\sigma_Y = 24KN/cm^2;\,k = 1.5}{a_{min,req} =?}$$

FBD

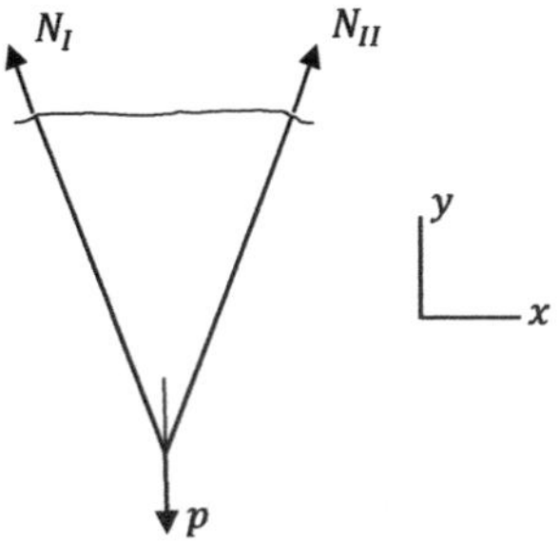

$$\sum F_x = 0 + N_I = N_{II}\;\sum F_y = 0\;2N_I\;\cos\alpha = P\;N_I = \frac{P}{2\cos\alpha} = \frac{P}{-\sqrt{2}};$$

$$a_I = a_{II} = a$$

$$\phi_I(x) = 1\;\phi_{II}(x) = 2$$

$$\sigma_I = \frac{P}{a\sqrt{2}} = 2\phi_{II} \max \frac{n(x)}{\phi(x)} = \frac{P}{\sqrt{2}} = \frac{120}{\sqrt{2}}$$

$$a_{min,req} = \frac{1}{16} \cdot \frac{120}{\sqrt{2}} = 5.303\, cm^2$$

Accordingly the probabilistic requirement (5.7)

$$R = P\left(\sum \leq \sigma_Y\right) \geq r$$

where σ_Y is a deterministic yield stress, gives rise to three basic problems of probabilistic strength of materials (PSM):

The first basic of PSM is checking the reliability. The probability distribution of the external loading, as well as the deterministic values of the cross-sectional area, the yield stress and the required reliability are all given, and it should be checked whether the actual reliability exceeds the required one.

Example 4.

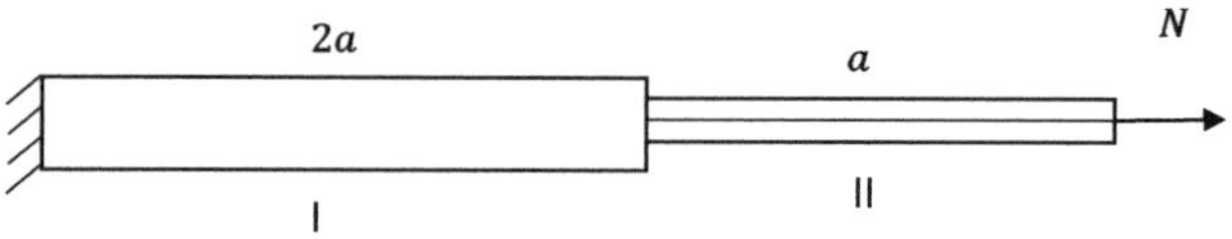

N is distributed exponentially with

$$F_N(n) = 1 - \exp\left[-\frac{n}{E(N)}\right]$$

with $E(N)$ given.

We have

$$R = P\left[\left(\sum I \leq \sigma_Y\right) \bigcap \left(\sum II \leq \sigma_Y\right)\right]$$

where $\sum I$ and $\sum II$ are the stresses in the two segments of the bar

$$R = P\left[\left(\frac{N}{2a} \leq \sigma_Y\right) \bigcap \left(\frac{N}{a} < \sigma_Y\right)\right] = P\left(\frac{N}{a} \leq \sigma_Y\right) = F_N\left(\sigma_Y a\right)$$

Since

$$\left\{\frac{N}{a} < \sigma_Y\right\} \bigcap \left\{\frac{N}{2a} \leq \sigma_Y\right\}$$

Therefore:

$$R = 1 - \exp\left[-\frac{\sigma_Y a}{E(N)}\right]$$

(See Eq. 5.15).

Suppose that $E(N)$ is proportional to $\sigma_Y a$, so that

$$E(N) = \alpha \sigma_Y a$$

then

$$R = 1 - e^{-\frac{1}{\alpha}}$$

Let the required reliability be $r = 0.999$; if $\alpha = \frac{1}{2}$, then the strength requirement is seen to be violated. If however $\alpha = 0.1$, then

$$R = 1 - e^{-10} = 0.9999596 > 0.999$$

and the strength requirement is met. In the first case the structure is unacceptable for use, in the second it is acceptable.

Generally, in our case the probabilistic strength requirement is met, if $1 - e^{-\frac{1}{\alpha}} \geq r$

or

$$\alpha < -\frac{1}{\ln(1-r)} = -\frac{1}{\ln 0.001}$$

Comparison with the deterministic strength requirement shows that the two types of requirement may generally yield different conclusions as to acceptability of the structure. Indeed, only in the case

$$-\frac{1}{\ln(1-2)} = 0.8333$$

or

$$r = 1 - e^{-1/2}$$

do we obtain the same limitation, $\alpha < 0.8333$ for the strength requirement to be met.

The second basic problem of PSM consists in determining maximum admissible values of the distribution parameters such that the probabilistic strength requirement is met.

Example 5.

The bar is subjected to a tensile force having a Rayleigh distribution (see Section 4.5):

$$f_N(n) = \begin{cases} 0, & n \le 0 \\ \frac{n}{b^2} e^{-n^2/2b^2}, & n \ge 0 \end{cases}$$

with parameter b^2, which is unknown. The cross sectional area of the bar is "a", and the required reliability is "r". We look for the maximum admissible value of "b" such that $R = (\sum < \sigma_Y) \ge r$.

Again we have

$$R = F_N(\sigma_Y a) = 1 - e^{-(\sigma_Y a)^2/2b^2}$$

The strength requirement is

$$1 - \exp[-(\sigma_Y a)^2/2b^2] \ge r$$

which yields

$$1 - r \ge \exp\left[\frac{-(\sigma_Y a)^2}{2b^2}\right]$$

$$\ln(1 - r) \ge \frac{-(\sigma_Y a)^2}{2b^2}$$

or

$$-\ln(1 - r) \le \frac{(\sigma_Y a)^2}{2b^2}$$

$$b^2 \le \frac{(\sigma_Y a)^2}{ln \frac{1}{1-r}}$$

The maximum admissible value of b is obtained by taking the equality sign in the equation

$$b_{max,adm} = \frac{\sigma_Y a}{\sqrt{ln \frac{1}{1-r}}}$$

So if $r = 0.999$, then $b_{max,adm} = 0.3805\, \sigma_Y a$.

It is of interest to check whether the mean stress is lower than the yield stress; or whether the strength requirement is met in the mean.

For a Rayleigh-distributed force,

$$E(N) = \frac{b\sqrt{\pi}}{\sqrt{2}} = 1.25b \text{ [see p. 75]}$$

Then

$$E(N) = \frac{1.25\sigma_Y a}{\sqrt{ln\frac{1}{1-r}}}$$

The mean stress is

$$E\left(\sum\right) = \frac{E(N)}{a} = \frac{1.25\sigma_Y}{\sqrt{ln\frac{1}{1-r}}}$$

For $r = 0.999$,

$$E\left(\sum\right) = 0.476\sigma_Y < \sigma_Y$$

and the strength requirement is met in the mean.

The third basic requirement of PSM is that of design; finding minimum required areas of the bar such that the required reliability is achieved.

Example 6.

The load has a Weibull distribution

$$F_N(n) = \begin{cases} 0, & n \leq 0 \\ \alpha\beta n^{\beta-1} e^{-\alpha n^{\beta}}, & n \geq 0 \end{cases}$$

The parameters α and β the required reliability r and the yield stress σ_Y are specified. Find the minimum required cross-sectional area.
Again

$$R = F_N(\sigma_Y a)$$

But

$$F_N(n) = 1 - e^{-\alpha n^{\beta}}$$

and

$$R = 1 - e^{-\alpha(\sigma_Y a)^{\beta}}$$

The strength requirement becomes

$$1 - e^{-\alpha(\sigma_y a)^{\beta}} \geq r$$

$a_{min,req}$ is given by

$$a_{min,req} = \frac{1}{\sigma_Y}\left(\frac{1}{\alpha}ln\frac{1}{1-r}\right)^{\frac{1}{\beta}}$$

If for example

$$\alpha = 2, \quad \beta = 3, \quad r = 0.999, \quad \sigma_Y = 24\,KN/cm^2$$

then

$$a_{min,req} = \frac{1}{24}\left(\frac{1}{2}ln\frac{1}{1-0.499}\right)^{\frac{1}{3}}$$

Remarks to Section 5.1

Equation (5.11) for the required cross-sectional area

$$a_{req} = \frac{E\,(N) + r(r-0.5)\sqrt{12}\sigma_N}{\sigma_Y} \tag{5.11}$$

was obtained for a uniformly distributed tensile force.

An analogous formula is obtainable for a normally distributed force. Indeed, according to Eq. (5.3)

$$R = F_N(\sigma_Y a) \geq r$$

where r is the required reliability. Now, if N is $N(E(N), \sigma_N^2)$ then

$$R = \frac{1}{2} + \text{erf}\left[\frac{\sigma_Y a - E(N)}{\sigma_N}\right] = r$$

The minimum permissible (required) cross-sectional area is obtained by using the equality sign in the last equation:

$$\frac{1}{2} + \text{erf}\left[\frac{\sigma_Y a - E\,(N)}{\sigma_N}\right] = r$$

or for a_{req}

$$a_{req} = \frac{E(N) + \sigma_N erf^{-1}\left(r - \frac{1}{2}\right)}{\sigma_Y}$$

if $\sigma_N \equiv 0$, i.e. N is a deterministic constant, we revert to a deterministic design:

$$a_{req} = \frac{E(N)}{\sigma_Y}$$

The probabilistic treatment "introduces" a new term $\sigma_N erf^{-1}(r - 1/2)$ in the nominator, and the table of error functions on p. 47 should be used to evaluate this contribution. We can use an additional table of values of the error function.

x	2.32	3.09	3.72	4.27	4.75
erf(x)	$\frac{1}{2} - 10^{-2}$	$\frac{1}{2} - 10^{-3}$	$\frac{1}{2} - 10^{-4}$	$\frac{1}{2} - 10^{-5}$	$\frac{1}{2} - 10^{-6}$
$\phi(x)$	$1 - 10^{-2}$	$1 - 10^{-3}$	$1 - 10^{-4}$	$1 - 10^{-5}$	$1 - 10^{-6}$
x	5.20	5.61	6.00	6.36	6.71
erf(x)	$\frac{1}{2} - 10^{-7}$	$\frac{1}{2} - 10^{-8}$	$\frac{1}{2} - 10^{-9}$	$\frac{1}{2} - 10^{-10}$	$\frac{1}{2} - 10^{-11}$
$\phi(x)$	$1 - 10^{-7}$	$1 - 10^{-8}$	$1 - 10^{-9}$	$1 - 10^{-10}$	$1 - 10^{-11}$

where

$$\Phi(x) = \frac{1}{\sqrt{2\pi}} \int_{-\infty}^{x} e^{-t^2/2} dt = \frac{1}{2} + \text{erf}(x)$$

for $r = 0.999$, $erf^{-1}(r - 1/2) = 3.09$ and

$$a_{req} = \frac{E(N) + 3.09\sigma_N}{\sigma_Y}$$

Now, if

$$E(N) \gg 3.09\sigma_N$$

deterministic analysis can again be used, otherwise probabilistic analysis is called for.

The table given hence shows that for $r \to 1$, $erf^{-1}(r - 1/2) \to \infty$, and a_{req} increases indefinitely, implying that the graph of a_{req} versus r has a vertical asymptote (see Figure below).

In this problem the question may arise whether it is justifiable to represent a tensile force by a normal random variable which can also take on negative values.

To overcome this difficult, we confine ourselves to a truncated normal distribution with positive values only. This is obtained from Equation (4.26) by setting

$x_1 = 0,\ x_2 \to \infty$, then

$$F_N(n) = \begin{cases} \dfrac{A}{\sigma\sqrt{2\pi}} \exp\left[-\dfrac{(x-no)^2}{2\sigma^2}\right], & 0 \le x \le \alpha \\ 0, & \text{otherwise} \end{cases}$$

from Equation (4.11) we find

$$A = \left[\frac{1}{2} + erf\left(\frac{no}{\sigma}\right)\right]^{-1}$$

and from Equation (4.28) the reliability requirement becomes

$$\frac{\mathrm{erf}\left(\frac{\sigma_x a - no}{6} + \mathrm{erf}\left(\frac{no}{\sigma}\right)\right.}{\frac{1}{2} + \mathrm{erf}\left(\frac{no}{\sigma}\right)} \ge r$$

which in turn yields a_{req}.

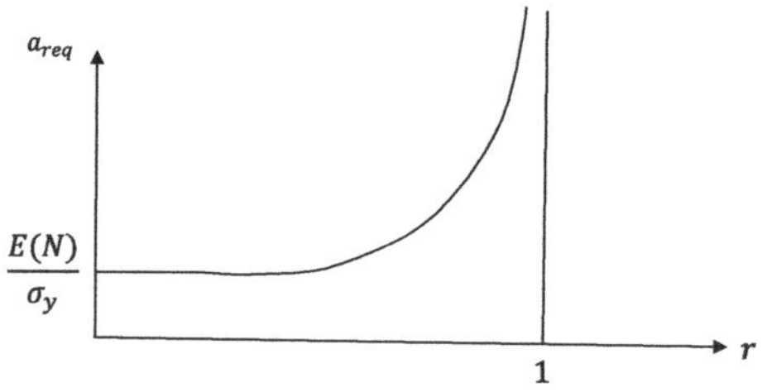

Remarks to Section 5.1 part 2

Equations (5.18)–(5.20) were obtained for a vanishing mean value of the force. Let us consider the case where $E(N) \neq 0$. Then, in Equation (5.17) the following distribution function (see Eq. 4.28, p. 85) has to be substituted for the truncated normal N:

$$F_N(n) = \begin{cases} 0, & -\infty \le n \le n_1 \\ \dfrac{\mathrm{erf}\left(\dfrac{n-n_0}{\sigma}\right) - \mathrm{erf}\left(\dfrac{n_1-n_0}{\sigma}\right)}{\mathrm{erf}\left(\dfrac{n_2-n_0}{\sigma}\right) - \mathrm{erf}\left(\dfrac{n_1-n_0}{\sigma}\right)}, & n_1 \le n \le n_2 \\ 1, & n_2 \le n < \infty \end{cases}$$

where n_0, n, and σ are parameters of the distribution Equation 5.17 becomes (by replacing σ_{allow} with σ_Y) :

$$\frac{\operatorname{erf}\left(\frac{\sigma_Y a-n_0}{\sigma}\right)-\operatorname{erf}\left(\frac{n_1-n_0}{\sigma}\right)}{\operatorname{erf}\left(\frac{n_2-n_0}{\sigma}\right)-\operatorname{erf}\left(\frac{n_1-n_0}{\sigma}\right)}$$

$$-\frac{\operatorname{erf}\left(\frac{\sigma_Y a-n_0}{\sigma}\right)-\operatorname{erf}\left(\frac{n_1-n_0}{\sigma}\right)}{\operatorname{erf}\left(\frac{n_2-n_0}{\sigma}\right)-\operatorname{erf}\left(\frac{n_1-n_0}{\sigma}\right)}\geq r$$

or

$$\frac{\operatorname{erf}\left(\frac{-\sigma_Y a-n_0}{\sigma}\right)-\operatorname{erf}\left(\frac{-\sigma_Y a-n_0}{\sigma}\right)}{\operatorname{erf}\left(\frac{n_2-n_0}{\sigma}\right)-\operatorname{erf}\left(\frac{n_1-n_0}{\sigma}\right)}\geq r$$

or due to the oddness of the error function:

$$\frac{\operatorname{erf}\left(\frac{\sigma_Y a-n_0}{\sigma}\right)+\operatorname{erf}\left(\frac{\sigma_Y a+n_0}{\sigma}\right)}{\operatorname{erf}\left(\frac{n_2-n_0}{\sigma}\right)-\operatorname{erf}\left(\frac{n_1-n_0}{\sigma}\right)}\geq r$$

Additional Examples to Section 6.7: "Approximate Evaluation of Moments of Functions"

Example 6.11 contrasted an exact and an approximate values of the mean of the quotient of two exponentially distributed random variables. It turned out that the exact value of $E(X|Y)$ tends to infinity, whereas, the approximation given by Equation 6.93.1 yields a finite number. Often, however, the contrast is "luckier". Some examples of either outcome follow.

Example 1

The radius R of the circle is a random variable with its probability density $f_R(r)$ given ($r \geq 0$). We are interested in the mean area $A = \pi R^2$ and its variance, both exactly and approximately.

The exact calculation yields

$$E(A) = E(\pi R^2) = \int_{-\infty}^{\infty} \pi r^2 f_R(r)dr = \int_{0}^{\infty} \pi r^2 f_R(r)dr = \pi E(R^2) = \pi M_2$$

since r cannot take on negative values; M_2 is the second moment. The approximate value is given by the first of the equations on page 209, namely,

$$E[g|X_1, X_2, \ldots, X_n) \simeq g[(X_1), E(X_2), \ldots, E(X_n)]$$

in our case $X_1 = R$, $g(R) = \pi R^2$, and

$$E(A) = E(\pi R^2) \simeq g[E(R)] = \pi [E(R)]^2 \equiv \tilde{E}(A)$$

i.e. it equals the area of a circle with the mean radius.

The difference of the exact and approximate values is

$$\begin{aligned} E(A) - \tilde{E}(A) &= \pi M_2 - \pi E^2(R) = \pi [M_2 - E^2(R)] \\ &= [(R^2) - E^2(R)] \end{aligned}$$

The expression in brackets is obviously $Var(R)$ (see Equation 3.33); that is

$$E(A) - \tilde{E}(A) = \pi Var(R)$$

so that, if $Var(R)$ is a small quantity,

$$(\pi Var(R) \ll \tilde{E}(A)), \text{ then}$$
$$E(A) \simeq \tilde{E}(A)$$

For numerical evaluation let R be an uniformly distributed random variable over (a, b). Then

$$E(A) = \pi E(R^2) = \pi \frac{b^3 - a^3}{3(b-a)} = \frac{\pi}{3}(b^2 + ab + a^2)$$

$$\tilde{E}(A) = \pi E^2(R) = \pi \frac{(a+b)^2}{4}$$

and

$$\begin{aligned} E(A) - \tilde{E}(A) &= \frac{\pi}{12}(4b^2 + 4ab + 4a^2 - 3a^2 - 6ab - 3b^2) \\ &= \frac{\pi}{12}(b^2 - 2ab - a^2) = \frac{\pi (b-a)^2}{12} \end{aligned}$$

as it should be, since

$$Var(R) = \frac{(b-a)^2}{12}$$

Now, if $a \to b$, then R tends to become a causally distributed random variable, with $Var(R) \to 0$ and

$$E(A) \to \tilde{E} \to \pi a^2$$

However the difference can be large if the variance of the radius is large, if $b \gg a$.

It is of interest to examine the correction term in Equation (6.91), for our scaler case this equation reads

$$E[g(X_1)] \simeq g[E(X_1)] + \frac{1}{2}\left(\frac{\partial^2 g}{\partial X_1^2}\right)\bigg|_{X_1=E(X_1)} Var(X_1)$$

which upon substitution of appropriate quantities, becomes

$$E(A) \simeq \pi E^2(R) + \frac{1}{2}(2\pi)Var(X_1)$$

since

$$g\,(X_1) = \pi X_1^2, \quad g' = 2\pi$$

which is then an exact formula.

Consider now the variance of the area:

$$\begin{aligned} Var(A) &= E(A^2) - E^2(A) \\ &= E(\pi^2 R^4) - E^2(\pi R^2) = \pi^2 E(R^4) - \pi^2 E^2(R^2) \end{aligned}$$

The approximate value is given by Equation (6.92), which becomes for the similar case:

$$\widetilde{Var}\,[g\,(X_1)] \simeq \left(\frac{\partial g}{\partial X_1}\right)^2 X_1 = E\,(X_1)\,Var(X_1) \tag{4}$$

or

$$\widetilde{Var}(A) = \; (2\pi r)|_{r=E(R)}\, Var(R) = 4\pi E^2(R) Var(R)$$

For uniformly distributed R is an interval (a, b) we again have

$$\begin{aligned} Var(A) &= \pi^2 E(R^4) - \pi^2 E^2(R^2) \\ &= \pi^2 \left[\frac{b^5 - a^5}{5(b-a)}\right] - \pi^2 \left[\frac{b^3 - a^3}{3(b-a)}\right]^2 \\ &= \pi^2 \frac{b^4 + ab^3 + a^2b^2 + a^3b + a^4}{5} - \pi^2 \left(\frac{b^2 + ab + a^2}{3}\right)^2 \\ &= \frac{\pi^2}{45}[9b^4 + 9ab^3 + 9a^2b^2 + 9a^3b + 9a^4 - 5b^4 - 5a^2 - 5a^2b^2 \\ &\quad -5a^4 - 10ab^3 - 10a^2b^2 - 10 - a^3b] \\ &= \frac{\pi^2}{45}(4b^4 - ab^3 - 6a^2b^2 - a^3b + 4a^4) \end{aligned}$$

whereas, the approximation yields

$$\widetilde{Var}(A) = 4\pi^2 \left(\frac{a+b}{2}\right)^2 \frac{(b-a)^2}{12}$$

For simplicity, let $a = 0$: then

$$Var(A) = \frac{\pi^2}{45}4b^4, \quad \widetilde{Var}(A) = \frac{\pi^2}{12}b^4$$

The attendant error is

$$n = \frac{Var(A) - \widetilde{Var}(A)}{Var(A)} \cdot 100\% = \frac{\frac{4}{45} - \frac{1}{12}}{\frac{4}{45}} 100\%$$
$$= \left(1 - \frac{45}{12 \cdot 4}\right) 100\% = \left(1 - \frac{45}{48}\right) 100\% = 6.25\%$$

Example 2

The maximum impact pressure of ocean waves on coastal structures may be determined by the formula given by A.H.S. Ang. And W.H. Tang (Probability Concepts in Engineering Planning and Design, Vol. 1, Basic Principles, p. 198):

$$P_{max} = 2 \cdot 7\frac{\rho K V^2}{d}$$

where ρ is the density of water, K- the length of the hypothetical piston, d- the thickness of the air cushion, and V- the horizontal velocity of the advancing wave.

We are interested in the mean of the peak impact pressure with ρ, K and d deterministic constants and V a random variable with mean $\bar{V}$ and $Var(V) = \sigma_V^2$.

The exact calculation yields

$$E(P_{max}) = 2 \cdot 7\frac{\rho K}{d}E(U^2) = 2 \cdot 7\frac{\rho K}{d}[Var(U) + E^2(U)]$$

The approximation is

$$E(P_{max}) = 2 \cdot 7\frac{\rho K}{d}E^2(U)$$

As is seen, the approximate formula is good for small values of $Var(U)$. The question arises, how to quantify this smallness. We calculate the percentagewise error:

$$\begin{aligned} n &= \frac{E(P_{max}) - \tilde{E}(P_{max})}{E(P_{max})} 100\% \\ &= \frac{[Var(U) + E^2(U)] - E^2(U)}{Var(U) + E^2(U)} 100\% \\ &= \frac{Var(U)}{Var(U) + E^2(U)} 100\% = \frac{100\%}{1 + \left[\frac{E(U)}{\sqrt{Var(U)}}\right]^2} \end{aligned}$$

Now we recall that

$$\frac{\sigma_U}{E(U)} = \gamma_U$$

is the coefficient of variation (See p. 65) of the random variable involved, with

$$n = \frac{100\%}{1 + (1/\gamma_u)^2}$$

so that for small values of the coefficient of variation $\gamma_U \ll 1$, $\gamma_U \gg 1$ and n will be a small quantity. If we allow a 5% error, then the approximate evaluation formulas can be used, if $\gamma_U \ll \gamma_U^*$ where the critical value γ_U^* satisfies the equation

$$5\% = \frac{100\%}{1 + \left(\frac{1}{\gamma_U}\right)^2}$$

yielding

$$Y_U^* = \frac{1}{\sqrt{19}} = 0.229$$

The coefficient of variation employed by Ang and Tang was 0.2 i.e. below the critical one, therefore, the attendant error of their calculation was less than 5%.

Example 3

Consider now a sphere with random radius R. We are interested in the mean volume of the sphere:

$$E(V) = E\left(\frac{4\pi^2}{3} R^3\right) = \frac{4\pi^2}{3} E(R^3)$$

The approximation used in the first equation on page 209 yields

$$\tilde{E}(V) = \frac{4\pi^2}{3} E^3(R)$$

again, for uniformly distributed R (we immediately set the interval $(a = 0, b)$). (See page 54):

$$E(V) = \frac{4\pi^2}{3}\frac{b^4}{4b} = \frac{\pi^2 b^3}{3}$$

$$\tilde{E}(V) = \frac{4\pi^2}{3} \cdot \left(\frac{b}{2}\right)^3 = \frac{\pi^2 b^3}{6}$$

i.e. yields only half the exact value. The refinement in Equation 6.91 yields

$$\tilde{E}(V) = \frac{4\pi^2}{3} E(R^3) + \frac{1}{2} \cdot (8\pi^2 r)\Big|_{r=R} Var(R)$$

$$= \frac{4}{3}\pi^2 \cdot E^3(R) + 4\pi^2 E(R) Var(R)$$

or, in our particular case

$$\tilde{E}(V) = \frac{4}{3}\pi^2 \cdot \left(\frac{b}{2}\right)^3 + 4\pi^2 \cdot \left(\frac{b}{2}\right)\left(\frac{b^2}{12}\right)$$

$$= \frac{\pi^2 b^3}{6} + \frac{\pi^2 b^3}{6} = \frac{\pi^2 b^3}{3}$$

again an exact value.

If, however, R is exponentially distributed then

$$\tilde{E}(V) = \frac{4\pi^2}{3} E(R^3) = \frac{4\pi^2}{3} \int_0^\infty r^3 a e^{-ar} dr$$

$$= \frac{4\pi^2}{3} a \left\{ e^{-ar} \left[\frac{r^3}{-a} - \frac{3r^2}{a^2} - \frac{6r}{a^3} - \frac{6}{a^4} \right]_0^\infty \right\} = \frac{4\pi^2}{3} a \cdot \frac{6}{a^4} = \frac{8\pi^2}{a^3}$$

whereas, the "rough" approximation yields

$$\tilde{E}(V) = \frac{4\pi^2}{3} E^3(R)$$

$$= \frac{4}{3}\frac{\pi^2}{a^3}$$

and the more "refined" approximation (Equation 6.91) yields (see also last equation on page 75)

$$\tilde{E}(V) \simeq \frac{4}{3}\pi^2 E^3(R) + 4\pi^2 \cdot E(R)Var(R)$$

$$= 4\pi^2\left[\frac{1}{3}\cdot\frac{1}{a^3} + \frac{1}{a}\cdot\frac{1}{a^2}\right] = 4\pi^2 \cdot \frac{4}{3a^3} = \frac{16\pi^2}{3a^3}$$

The error now yielded by the "rough" approximation is

$$n_1 = \frac{8-\frac{4}{3}}{8}100\% = \left(1 - \frac{4}{3\cdot 8}\right)\cdot 100\% = \frac{7}{6}\cdot 100\% > 116\%$$

whereas that associated with the more refined approximation is

$$n_2 = \frac{8-\frac{16}{3}}{8}100\% = \left(1-\frac{2}{3}\right)100\% = 33.3\%$$

which is still high. Indeed the input random variable, in this case an exponential one, has a coefficient of variation actually equal to unity:

$$\gamma_R = \frac{\sigma R}{E(R)} = \frac{\sqrt{Var(R)}}{E(R)} = \frac{1/a}{1/a} = 1$$

Example 4

A bar is subjected to a force of 1 KN, over an area X is a random variable with $f_X(x)$ given $X \geq 0$. We are interested in the average stress in the bar cross-section.

Now:

$$E\left(\sum\right) = E\left(\frac{1}{X}\right) = \int_0^\infty \frac{1}{x} f_X(x)dx$$

the approximate formula yields

$$\tilde{E}\left(\sum\right) = E\left(\frac{1}{X}\right) \cong \frac{1}{E(X)}$$

i.e. stress is in a bar with the average cross-sectional area.

To calculate the error involved, we specify $f_X(x)$. Let X be a uniform random variable over the interval $(\alpha, \alpha + \beta)$. Then

$$f_X(x) = \begin{cases} \frac{1}{\beta}, & \text{for } x \epsilon [\alpha, \alpha + \beta] \\ 0, & \text{otherwise} \end{cases}$$

and

$$E\left(\sum\right) = \frac{1}{\beta} \int_{\alpha}^{\alpha+\beta} \frac{1}{x} dx = \frac{1}{\beta} \ln x \Big]_{\alpha}^{\alpha+\beta} = \frac{1}{\beta} \ln \frac{\alpha + \beta}{\alpha}$$

and is finite for $\alpha > 0$, and infinity for $\alpha = 0$. Yet in both cases

$$\tilde{E}\left(\sum\right) = \frac{1}{E(X)} = \frac{1}{(\alpha + \alpha + \beta/2)} = \frac{1}{\alpha + \beta/2}$$

For small β/α we anticipate small differences. In fact,

$$\gamma_X = \frac{\sigma_x}{E(X)} = \frac{\sqrt{(\alpha + \beta - \alpha)^2/12}}{(\alpha + \alpha + \beta)/2} = \frac{\beta/\sqrt{12}}{\alpha + \beta/2} = \frac{\beta/\alpha}{\sqrt{12}(1 + 0.5\beta/\alpha)}$$

is again small.

Indeed, expansion of the exact formula yields

$$E\left(\sum\right) = \frac{1}{\beta} \ln\left(1 + \frac{\beta}{\alpha}\right) = \frac{1}{\beta} \cdot \frac{\beta}{\alpha} = \frac{1}{\alpha}$$

which is close to $\tilde{E}(\sum)$ for $\beta \ll \alpha$.

Now, if γ_X is not small, errors will be larger. If X is exponentially distributed then

$$E\left(\sum\right) = \int_0^{\infty} \frac{1}{x} [a e^{-ax}] dx = a \int_0^{\infty} \frac{1}{y} e^{-y} dy \to \infty$$

this integral is known to diverge. By contrast

$$E\left(\sum\right) = \frac{1}{E(X)} = a$$

The example is close in character to the Example 6.11.

The conclusion to be drawn from them is that approximate moments may be satisfactorily provided the coefficients of variation of the random variables in question are sufficiently small, or if the functional relationships involved are close to linear.

Example 5.

The frequency of a single-degree-of-freedom system given by

$$\Omega = \sqrt{\frac{K}{M}}$$

where K is spring stiffness, M is mass. Determine approximately the mean and mean square value of the frequency, if $E(K)$ and $E(M)$ are given and $\gamma_K = 0.2$, $\gamma_M = 0.1$. We have

$$\tilde{E}(\Omega) = \sqrt{\frac{E(K)}{E(M)}}$$

$$\frac{\partial \Omega}{\partial K} = \left(\frac{1}{\sqrt{M}} \cdot \frac{1}{2}\frac{1}{\sqrt{K}}\right)\Bigg|_{\substack{M = E(M)\\ K = E(K)}} = \frac{1}{2\sqrt{E\,(M)\,E(K))}}$$

$$\frac{\partial \Omega}{\partial M} = \sqrt{K} \cdot \left(-\frac{1}{2}M^{-\frac{3}{2}}\right)\Bigg|_{\substack{M = E(M)\\ K = E(K)}} = -\frac{1}{2}\frac{\sqrt{E(K)}}{E(M)\sqrt{E(M)}}$$

$$Var(\Omega) = \left(\frac{\partial \Omega}{\partial K}\right)^2\Bigg|_{\substack{M = E(M)\\ K = E(K)}} Var(K) + \left(\frac{\partial \Omega}{\partial M}\right)\Bigg|_{\substack{M = E(M)\\ K = E(K)}} Var(M)$$

$$= \frac{1}{4E(M)E(K)} \cdot Var(K) + \frac{1}{4}\frac{E(K)}{E^3(M)} Var(M)$$

$$Var(\Omega) = \tilde{E}^2(\Omega) \cdot \left[\frac{1}{4}\frac{Var(K)}{E^2(K)} + \frac{1}{4}\frac{Var(M)}{E^2(M)}\right]$$

$$= \frac{1}{4}\tilde{E}^2(\Omega)(\gamma_K^2 + \gamma_M^2) = E^2(\Omega)\frac{1}{4}(0.2^2 + 0.1^2)$$

$$= \tilde{E}^2(\Omega)\frac{0.04 + 0.01}{4} = 0.0125\tilde{E}^2(\Omega)$$

$$\tilde{\gamma}_\Omega = \frac{\sqrt{\widetilde{Var}(\Omega)}}{\tilde{E}(\Omega)} = \sqrt{0.0125} = 0.112$$

In case stiffness and mass have the same coefficients of variation we get

$$\tilde{\gamma}_\Omega = \frac{\sqrt{\gamma_K^2 + \gamma_M^2}}{4} = \frac{\sqrt{2}}{2}\gamma_K = 0.71\gamma_K$$

i.e. the coefficient of variation of the natural frequency is less than that of the mass or the stiffness.

Additional Material to Chapter 6

In Example 6.12 we discussed a "ball-urn" example illustrating the consequences of omission of the correlations of random variables. Here is another, more practical problem:

What is the mathematical expectation and variance of the number of hits occurring in n dependent or independent firing sequences with hit probabilities $P_1, P_2, \ldots P_n$?

We consider the random variable X denoting the number of hits

$$X = X_1 + X_2 + X_n$$

where X_i is the number of hits in its firing sequence. Moreover,

$$X_i = \begin{cases} 1, & \text{if } j\text{th trial is a success} \\ 0, & \text{if } j\text{th trial is a failure} \end{cases}$$

hence X_i is a random variable with mathematical expectation

$$E(X_i) = 1 \cdot P_i + 0 \cdot (1 - P_i) = P_i$$

Now, by virtue of Equation (6.76)

$$\begin{aligned} E(X) &= E(X_i + X_2 + \cdots + X_n) \\ &= E(X_i) + E(X_2) + \cdots + E(X_n) \end{aligned}$$

irrespective of whether X_i are dependent or independent.

Now for the variance

$$Var(X_i) = (0 - P_i)^2(1 - P_i) + (1 - P_i)^2 P_i = P_i(1 - P_i)$$

for $Var(X)$ we obtain

$$\begin{aligned} Var(X) &= Var(X_i + X_2 + \cdots + X_n) \\ &= \sum_{j=1}^{n} Var(X_i) + \sum_{j=1_{j\neq k}}^{n} \sum_{k=1}^{n} Cov(X_j, X_k) \end{aligned}$$

due to Equation (6.77).

This equation can be put in the form

$$Var(X) = \sum_{j=1}^{n} Var(X_i) + 2 \sum_{j=1_{j<k}}^{n} \sum_{j=k}^{n} Cov(X_j, X_k)$$

if the sequences are independent, then $Cov(X_j, X_k) = 0$ and

$$Var(X) = \sum_{j=1}^{n} Var(X_j).$$

Consider the case of dependent firing sequences. In accordance with Equation 6.69

$$\begin{aligned} Cov(X_j, X_k) &= E(X_j, X_k) - E(X_j)E(X_k) \\ &= E(X_j, X_k) - P_j P_k \end{aligned}$$

Consider now the random variable $\{X_j \cap X_k\}$ the intersection of the random variables X_j and X_k. It takes on the value 0 if one of the sequences is a failure and unity of both are successful. Accordingly, $E(X_j X_k)$ equals the probability of success at both sequences, which we denote by P_{jk}:

$$P_{jk} = Prob[(X_j = 1) \cap (X_k = 1)]$$

Then

$$\begin{aligned} E(X_j X_k) &= P_{jk} \\ Cov(X_j, X_k) &= P_{jk} - P_k P_k \end{aligned}$$

which in turn yields

$$Var(X) = \sum_{j=1}^{n} P_j(1 - P_j) + 2 \sum_{j=1_{j<k}}^{n} \sum_{k=1}^{n} (P_{jk} - P_j P_k)$$

the first term representing the variance of the number of hits in independent sequences, and the second the correction for dependence.

Let

$$P_1 = P_2 \ldots = P_n = P$$

with the sequences dependent and all hits concentrated in a single bull's eye. This is a simplified case of volley firing when the volley either hits or misses as a single entity. Thus a hit by one shot means hits by all the others, and:

$$P_{jk} = P_j P_k$$

which yields

$$Var(X) = np(1-P) + 2\frac{n(n-1)}{2}(P - P^2)$$
$$= nP(1-P) + n(n-1)P(1-P) = n^2P(1-P)$$

with the covariances disregarded, we have the approximation (which is exact for independent sequences)

$$\widetilde{Var}(kX) = nP(1-P)$$

The attendant error is

$$n = \frac{Var(X) - \widetilde{Var}(X)}{Var(X)}100\% = \frac{n(n-1)P(1-P)}{n^{2P}(1-P)}100\% = \frac{(n-1)}{n}100\%$$

namely, 50% for $n = 2$, 66.7% for $n = 3$ and 75% $n = 4$. The error tends to 100% for $n \to \infty$.

Additional Material to Sec. 7.4

The Central Limit Theorem and Reliability Estimate

Say that we have a problem of probabilistic design of a structural member under an axial force with chi-squared distribution (extension of Problem 5.1).

Find "a", such that $R > r = 0.995$, $\sigma_Y = 22\,kN/mm$, $\nu = 35$.

Table 5.1 cannot be used for finding the value $\chi^2\alpha, \nu$ corresponding to $\alpha = 1 - R = 0.005$, since it contains only values up to $\nu = 30$. However, as was shown in Example 6.13, if $X_j(j = 1, 2, \ldots, 2\nu)$ are independent random variables, identically normally distributed, $N(a\sigma^2)$ — then

$$\frac{Y}{\sigma^2} = \frac{1}{\sigma^2}\sum_{j=1}^{2\nu}(X_d - a)^2$$

has a chi-square distribution.

As the sum of independent random variables each of which is square of a standard normal variable, has a chi-square distribution, we can use the central limit theorem to approximate the exact probability law of N

Now, since N has a chi-square distribution; according to the central limit theorem

$$P(N \le n) \simeq \frac{1}{2} + \text{erf}\left(\frac{n - E(N)}{\sigma_N}\right)$$

But (see sec. 4.7):

$$E(N) = \nu, \quad \sigma_N^2 = 2\nu$$

Since $m = 2\nu$ in the notation of problem 5.1

Therefore

$$P(N \leq n) \simeq \frac{1}{2} + \text{erf}\left(\frac{n-\nu}{\sqrt{2\nu}}\right)$$

the reliability is then

$$R = P(N < \sigma_Y A) \simeq \frac{1}{2} + \text{erf}\left(\frac{\sigma_y a - \nu}{\sqrt{2\nu}}\right) \geq r$$

for a_{req} we have

$$\frac{1}{2} + \text{erf}\left(\frac{\sigma_y a_{req} - \nu}{\sqrt{2\nu}}\right) = r$$

and

$$\sigma_y a_{req} = \nu + \sqrt{2\nu}\text{erf}^{-1}\left(r - \frac{1}{2}\right)$$

Alternatively, this implies approximation of the r-th quantitle $\chi^2_{1-r,\nu}$ of chi-square distribution:

$$\chi^2_{1-r,\nu} \simeq \nu + \sqrt{2\nu}erf^{-1}\left(r - \frac{1}{2}\right)$$

for $r = 0.995$, $erf^{-1}(0.495) = 2.58$ (see p. 471), and

$$a_{req} = \frac{\nu + 2.68\sqrt{2\nu}}{\sigma_Y}$$

for $\nu = 35$,

$$a_{req} = \frac{57.422}{22} = 2.61mm^2$$

for the design problem solved in Problem 5.1 for $\nu = 16$, use of the present approximation would yield

$$\chi^2_{0.005,16} = 16 + \sqrt{32} \cdot 2.68 = 31.160$$

Interestingly the exact value listed Table 5.1 is 34.267.

Note that Sir R.A. Fisher delivered an even more accurate approximation [see eq. H.J. Larson, "Introduction to Probability Theory and Statistical Inference",

Third Edition, John Wiley and Sons, New York, 1982, p. 293] for $\chi^2\alpha, \nu$ of which reads for our problem:

$$\chi^2_{1-r,\nu} = \frac{\left[erf^{-1}\left(r - \frac{1}{2}\right) + \sqrt{2\nu - 1}\right]^2}{2}$$

For the case discussed, $\nu = 35, r = 0.995$ we find

$$\chi^2_{0.005,35} \simeq 60.353$$

instead of 57.422 according to the previous approximation. Under these approximation, $a_{req} = 2.79mm^2$ and not 2.61 mm^2.

For the case $\nu = 16$ Fisher's approximation yields, $\chi^2 0.005, 16 = 34.013$ which is close to that given in Table 5.1, namely 34.267.

Comments on Example 7.11

Let us prove that the maximum moment

$$M_{max} = M_1 + M_2 + \cdots + M_n$$

has a gamma distribution, if the constituent moments have an exponential distribution. Indeed, for

$$f_{M_j}(m_j) = \lambda_j e^{-\lambda_j m_j}, \quad m_j > 0$$

we have the characteristic function (see p. 75)

$$M_{m_j}(\theta) = \frac{\lambda_j}{\lambda_j - i\theta}$$

and, since M_j's are independent, we get, via Equation (6.100)

$$M_{M_{max}} = (\theta) = \prod_{j=1}^{n} \frac{\lambda_j}{\lambda_j - i\theta}$$

Now, if all $\lambda_j \equiv \lambda$, we get

$$M_{M_{max}} = \prod_{j=1}^{n} \frac{\lambda}{\lambda - i\theta} = \frac{\lambda^n}{(\lambda - i\theta)^n} = \frac{1}{(1 - i\theta/\lambda)^n}$$

Comparison with the characteristic function of the gamma random variable (p. 76) reveals that M_{max} has gamma distribution, with $\beta = 1/\lambda$ and $\alpha = n - 1$.

Now we will derive the probability distribution function (given in Equation 7.68) of a gamma random variable $M_{max} \equiv M$, with

$$f_M(m) = \frac{\lambda(\lambda m)^{n-1}}{(n-1)!} e^{-\lambda m}$$

with $\lambda = 1/\beta, \alpha = n - 1$ in Equation 4.10, defining the gamma random variable.

In fact

$$F_M(m) = Prob(M \le m) = 1 - Prob(M > m)$$
$$= 1 - \int_m^\infty \frac{\lambda}{(n-1)!}(\lambda y)^{n-1} e^{-\lambda y} dy, \quad m > 0$$

We let

$$\lambda y = z, \quad dy = dz/\lambda$$

yielding

$$F_M(m) = 1 - \int_{\lambda m}^\infty e^{-z} z^{n-1} dz$$

We integrate by parts, letting

$$u = z^{n-1}, \quad dv = e^{-z} dz$$

This yields

$$du = (n-1)z^{n-2}, \quad v = -e^{-z}$$

Hence

$$(n-1)!I = -z^{n-1}e^{-z}]_{\lambda m}^\infty + \int_{\lambda m}^\infty e^{-z}(n-1)z^{n-2}dz$$
$$= (\lambda m)^{n-1}e^{-\lambda m} + (n-1)\int_{\lambda m}^\infty e^{-z} z^{n-2} dz$$

The integral in this expression coincides with the original integral with n replaced by $n-1$. Thus continuing to integrate by parts we obtain, since n is a positive integer:

$$(n-1)!I = e^{-\lambda m}[(\lambda m)^{n-1} + (n-1)(\lambda m)^{n-2}$$
$$+(n-1)(n-2)(\lambda m)^{n-3} + \cdots + (n-1)!]$$

Therefore

$$I = e^{-\lambda m}\left[1 + (\lambda m) + \frac{(\lambda m)^2}{2!} + \cdots + \frac{(\lambda m)^{n-1}}{(n-1)!}\right]$$

which proves Equation (7.68).

Interestingly $F_M(n)$ can be put in the form

$$F_M(m) = 1 - \sum_{j=0}^{n-1} \frac{e^{-\lambda m}(\lambda m)^j}{j!}, \quad m > 0$$

implying that the probability distribution function of the gamma random variable may be expressed in terms of tabulated (tables are available in literature) probability distributions of Poisson distribution. With $\lambda m = a$ in Equation 4.6, (replacing $m \to r$ in Equation 4.6)

$$P(X = r) = \frac{a^r}{r!}e^{-a}, \quad r = 1, 2, \ldots$$

and the probability distribution function is

$$F_x(k) = Prob(x \leq k) = \sum_{r=0}^{k} P(X = r) = \sum_{r=0}^{k} \frac{a^r}{r!}e^{-a}.$$

Remark on Section 11.2

Section 11.2 describes the congruential method of deriving an uniformly distributed random variable; either

$$x_i = ax_{i-1}(\text{mod}\, m)$$

or

$$x_i = (ax_{i-1} + b)(\text{mod}\, m)$$

where a is the multiplier, B is the increment, and m-the modulus, and are nonnegative numbers. The modulus notation $(\text{mod}\, m)$ means that

$$x_i = ax_{i-1} + b - mk_{i-1}$$

where

$$k_{i-1} = \left[\frac{ax_{i-1} + b}{m}\right]$$

denotes the largest positive integer in $(ax_{i-1} + B)/m$.

For example let $a = b = x_0 = 3$, and $m = 5$. Then

$$x_i = 3, 2, 4, 0, 3.$$